ADVANCED POWER SOURCES FOR SPACE MISSIONS

Committee on Advanced Space Based High Power Technologies
Energy Engineering Board
Commission on Engineering and Technical Systems
National Research Council

NATIONAL ACADEMY PRESS
Washington, D.C. 1989

National Academy Press • 2101 Constitution Avenue, N.W. • Washington, D. C. 20418

NOTICE: The project that is the subject of this report was approved by the Governing Board of the National Research Council, whose members are drawn from the councils of the National Academy of Sciences, the National Academy of Engineering, and the Institute of Medicine. The members of the committee responsible for the report were chosen for their special competences and with regard for appropriate balance.

This report has been reviewed by a group other than the authors according to procedures approved by a Report Review Committee consisting of members of the National Academy of Sciences, the National Academy of Engineering, and the Institute of Medicine.

This report and the study on which it was based were supported by Contract No. F49620-85-C-107 from the United States Air Force to the National Academy of Sciences.

Library of Congress Catalog Card No. 88-63907
ISBN 0-309-03999-1

Cover art: Robert T. McCall.

The cover depicts an artist's conception of a directed-energy space weapon, based on suggestions provided by scientists and engineers working on advanced space weapons. Such a weapon might form part of a U.S. strategic defense system to be used against nuclear missiles. The directed-energy weapon shown at the lower part of the spacecraft is a free-electron laser emitting laser beams downward toward an enemy target. Other related concepts that may eventually be part of an actual weapons system in space are also shown. The steady "housekeeping" power needed to operate the weapons platform in peacetime is provided by an SP-100 space nuclear reactor, in the upper part of the painting. Here the "battle mode" power is produced chemically by the reaction of hydrogen with oxygen, which produces high-pressure steam to drive turbogenerators. The resulting steam effluent is then released into space, as shown, during battle. The output optical sensors are protected by large clamshell doors until actual operation. Small kinetic-energy vehicles are shown being released as a defense against antisatellite missiles. Actual Strategic Defense Initiative weapons platforms may not closely resemble the system presented in this painting, in which Robert McCall has endeavored to capture the power and essence of a potential space defense system.

COMMITTEE ON ADVANCED SPACE BASED HIGH POWER TECHNOLOGIES

JOSEPH G. GAVIN, JR. (*Chairman*), Grumman Corporation, Bethpage, New York
TOMMY R. BURKES, Texas Tech University
ROBERT E. ENGLISH, Consultant, Cleveland, Ohio
NICHOLAS J. GRANT, Massachusetts Institute of Technology
GERALD L. KULCINSKI, University of Wisconsin, Madison
JEROME P. MULLIN, Sundstrand Corporation, Rockford, Illinois
K. LEE PEDDICORD, Texas A&M University
CAROLYN K. PURVIS, NASA Lewis Research Center, Cleveland, Ohio
W. JAMES SARJEANT, State University of New York at Buffalo
J. PACE VANDEVENDER, Sandia National Laboratories, Albuquerque, New Mexico

Energy Engineering Board Liaison

S. WILLIAM GOUSE, The Mitre Corporation, McLean, Virginia

Technical Adviser

Z. J. JOHN STEKLY, Intermagnetics General Corporation, Acton, Massachusetts

Staff

ARCHIE L. WOOD, *Director*, Energy Engineering Board
ROBERT COHEN, *Study Director*
ANN M. STARK, *Study Assistant*

ENERGY ENGINEERING BOARD

Technical Advisory Panel

HAROLD M. AGNEW, GA Technologies, Inc., Solana Beach, California
FLOYD L. CULLER, JR., Electric Power Research Institute, Palo Alto, California
KENT F. HANSEN, Massachusetts Institute of Technology
MILTON PIKARSKY, The City College, New York, New York
CHAUNCEY STARR, Electric Power Research Institute, Palo Alto, California
HERBERT H. WOODSON, The University of Texas at Austin

Staff

ARCHIE L. WOOD, *Director*
DRUSILLA BARNES, *Administrative Secretary*
ROBERT COHEN, *Senior Program Officer*
FREDERIC MARCH, *Senior Program Officer*
CARLITA M. PERRY, *Administrative Associate*
ROSENA A. RICKS, *Administrative Secretary*
ANN M. STARK, *Administrative Assistant*
JAMES J. ZUCCHETTO, *Senior Program Officer*

Preface

This study, conducted under the auspices of the Energy Engineering Board of the National Research Council, examines the status of and outlook for advanced power sources for space missions. The study resulted from a request by the U.S. Department of Defense (DOD) for an independent review relating to the space power requirements of its Strategic Defense Initiative (SDI).

Initial impetus for the study came from the U.S. Air Force Wright Aeronautical Laboratories, at about the time the SDI Organization (SDIO) was being formed in DOD. Initially, the charge to this committee included these tasks:

- Evaluate the planning for the development of advanced space-based high-power technologies to determine the best combination of technology options that should be pursued.
- Critique current SDI power development plans and objectives.
- Identify an alternate power program plan that would meet SDI requirements for space-based power.
- Identify technology development approaches that could lead to enabling power system capabilities for future space-based defensive systems.

To examine the relevant but less demanding power needs of other U.S. space missions, the scope of the study was subsequently broadened to include consideration of military space power requirements other than those of SDIO and of potential civil space power requirements, especially those of the National Aeronautics and Space Administration (NASA), where power will be needed for earth-orbital, interplanetary, and lunar-surface missions.

A forerunner to this study, with emphasis on space nuclear power, was conducted by the National Research Council's Committee on Advanced Nuclear Systems, under the chairmanship of John M. Deutch. That study led to the report, "Advanced Nuclear Systems for Portable Power in Space," published in 1983.

In accordance with its charter, this committee has taken a broad look at candidate power technologies for space missions, both civil and military. At the same time, special emphasis was given to studying the specific space power requirements of the SDI program, and possible programmatic courses of action for satisfying them. In the study, technology options were mainly considered for their capability to provide space-based power for applications other than propulsion.

On behalf of SDIO, Richard Verga, Robert L. Wiley, David Buden, and Richard G. Honneywell provided useful inputs and cooperation throughout the study. Richard G. Honneywell, of Air Force Wright Aeronautical Laboratories, initiated the request and contract for the study. By the time the committee began its work, the focal point for government technical interaction with the committee had shifted to the SDIO Power Program Office, headed by Richard Verga. The committee received timely, useful briefings and valuable written material from that office, its contractors, and other individuals. David Buden served as a committee member for several months, at which time he resigned after accepting an offer to become Richard Verga's deputy. Louis O. Cropp and his colleagues at Sandia National Laboratories furnished the committee with numerous technical inputs and publications. Phillip N. Mace and Milan Nikolich, of W. J. Schafer Associates, Inc., frequently provided technical and logistical assistance to the committee in that company's capacity as a support-services contractor to SDIO.

Arrangements to conduct the study were facilitated by Dennis F. Miller, Director of the Energy Engineering Board until November 1987. He was succeeded by Archie L. Wood in December 1987. Robert Cohen served as Study Director and as Editor of this report.

JOSEPH G. GAVIN, JR., *Chairman*
Committee on Advanced Space Based High Power Technologies

Contents

ADVANCED POWER SOURCES FOR SPACE MISSIONS

Executive Summary

This study focuses on approaches to satisfying the power requirements of space-based Strategic Defense Initiative (SDI) missions. The study also considers the power requirements for non-SDI military space missions and for civil space missions of the National Aeronautics and Space Administration (NASA). The more demanding SDI power requirements appear to encompass many, if not all, of the power requirements for those missions. Study results indicate that practical fulfillment of SDI requirements will necessitate substantial advances in the state of the art of power technology.

SDI goals include the capability to operate space-based beam weapons, sometimes referred to as directed-energy weapons. Such weapons pose unprecedented power requirements, both during preparation for battle and during battle conditions. The power regimes for those two sets of applications are referred to as alert mode and burst mode, respectively.

Alert-mode power requirements are broadly stated, inadequately defined, and still evolving. They are presently stated to range from about 100 kw to a few megawatts for cumulative durations of about a year or more. These power and time parameters correspond to an energy (power multiplied by time) requirement in space ranging from about a million kilowatt-hours to several billion kilowatt-hours. Burst-mode power requirements are roughly estimated to range from tens to hundreds of megawatts for durations of a few hundred to a few thousand seconds, corresponding to space-based energy requirements ranging from hundreds to millions of kilowatt-hours.

Complete study findings, conclusions, and recommendations are contained in the body of this report and are compiled in Appendix E. This summary restates all of the study recommendations and highlights the most significant study conclusions that led to them. Those conclusions are that:

- There are two likely energy sources, chemical and nuclear, for powering SDI directed-energy weapons during the alert and burst modes. The choice between chemical and nuclear space power systems depends in large part on the total duration during which power must be provided. On the basis of mass-effectiveness, large durations favor the nuclear reactor space power system and short durations favor chemical power systems if their effluents can be tolerated. For alert-mode requirements at the low-power end of the requirement range stated above, a solar space power system might qualify.
- Multimegawatt space power sources appear to be a necessity for the burst mode.
- Pending resolution of effluent tolerability, open-cycle power systems (systems whose working fluid is used only once) appear to be the most mass-effective solution to burst-mode electrical power needs in the multimegawatt regime. If an open-cycle system cannot be developed, or if its interactions with the spacecraft, weapons, and sensors prove unacceptable, the entire SDI concept will be severely penalized from the standpoints of cost and launch weight.
- A nuclear reactor power system may prove to be the only viable option for powering the SDI burst mode (if effluents from chemical power sources prove to be intolerable) and for powering the SDI alert mode (if the total energy requirements of the alert mode exceed what can be provided by chemical or solar means).
- A space nuclear reactor power system, such as the SP-100 system presently being developed, would be a step toward meeting SDI requirements and would be applicable to other civil and military space missions. Early deployment of an experimental space power system, possibly the SP-100 system, would be useful to provide confirmation of design assumptions prior to commitment to an SDI system.
- Beaming power upward from earth by microwaves or lasers has not been extensively explored as a space power option, but may be worthy of further study.

• Estimated gross masses of SDI space power systems analyzed in existing studies appear unacceptably large to operate major space-based weapons. At these projected masses, the feasibility of space power systems needed for high-power SDI concepts appears impractical from both cost and launch considerations.

• At the current rate of power technology development, power systems appear to be a pacing item for the successful development of SDI directed-energy weapon systems. Accordingly, either major innovations in power systems and power system components will be required or SDI power requirements will have to be relaxed.

• Existing SDI space power architecture system studies do not provide an adequate basis for evaluating or comparing cost or cost-effectiveness among the space power systems examined, do not adequately address questions of survivability, reliability, maintainability, and operational readiness, and do not adequately relate to the design of complete SDI spacecraft systems.

The committee arrived at the following recommendations:

Recommendation 1 (Chapter 4): Using the latest available information, an in-depth, full-vehicle-system preliminary design study—for two substantially different candidate power systems for a common weapon platform—should be performed now, in order to reveal secondary or tertiary requirements and limitations in the technology base that are not readily apparent in the current space power architecture system studies. Care should be exercised in establishing viable technical assumptions and performance requirements, including survivability, maintainability, availability, ramp-rate, voltage, current, torque, effluents, and so on. This study should carefully define the available technologies, their deficiencies, and high-leverage areas where investment will produce significant improvement. The requirement for both alert-mode and burst-mode power and energy must be better defined. Such an in-depth system study will improve the basis for power system selection, and could also be helpful in refining mission requirements.

Recommendation 2 (Chapter 3): To remove a major obstacle to achieving SDI burst-mode objectives, estimate as soon as practicable the tolerable on-orbit concentrations of effluents. These estimates should be based—to the maximum extent possible—on the results of space experiments, and should take into account impacts of effluents on high-voltage insulation, space-platform sensors and weapons, the orbital environment, and power generation and distribution.

Recommendation 3: Rearrange space power R&D priorities as follows:

a. **(Chapter 3) Give early, careful consideration to the regulatory, safety, and National Environmental Policy Act requirements for space nuclear power systems from manufacture through launch, orbital service, safe-orbit requirements, and disposition.**
b. **(Chapter 6) The SP-100 nuclear power system is applicable both to SDI requirements and to other civil and military space missions. Therefore, SP-100 development should be completed, following critical reviews of SP-100 performance goals, design, and design margins.**
c. **(Chapter 6) SDI burst-mode requirements exceed by one or more orders of magnitude the maximum power output of the SP-100. Therefore, both the nuclear and nonnuclear SDI multimegawatt programs should be pursued.** Hardware development should be coordinated with the results of implementing Recommendation 5.

Recommendation 4 (Chapter 6): Consider deploying the SP-100 or a chemical power system on an unmanned orbital platform at an early date. Such an orbital "wall socket" could power a number of scientific and engineering experiments. It would concurrently provide experience relevant to practical operation of a space power system similar to systems that might be required by the SDI alert and burst modes.

Recommendation 5 (Chapter 5): Make additional and effective investments now in technology and demonstrations leading to advanced components, including but not limited to:

- **thermal management, including radiators;**
- **materials: structural, thermal, environmental and superconducting;**
- **electrical generation, conditioning, switching, transmission, and storage; and**
- **long-term cryostorage of H_2 and O_2.**

Advances in these areas will reduce power system mass and

environmental impacts, improve system reliability, and, in the long term, reduce life-cycle power system cost.

Recommendation 6 (Chapter 3): **Review again the potential for ground-based power generation (or energy storage) with subsequent electromagnetic transmission to orbit.**

Recommendation 7 (Chapter 2): **After adequate evaluation of potential threats, further analyze the subject of vulnerability and survivability, mainly at the overall system level.** Data resulting from implementing Recommendation 1 would be appropriate for this further analysis. Pending such analysis, candidate power systems should be screened for their potential to satisfy interim SDI Organization (SDIO) survivability requirements, reserving judgment as to when or whether those requirements should constrain technology development. Convey the screening results to the advocates of those candidate power systems, to stimulate their finding ways to enhance survivability as they develop the technology.

Recommendation 8 (Chapter 6): **To further U.S. capabilities and progress in civil as well as military applications of power technology, both on the ground and in space, and to maintain a rate of progress in advanced technologies adequate to satisfy national needs for space power, plan and implement a focused federal program to develop the requisite space power technologies and systems. This program—based on a multiyear federal commitment—should be at least as large as the present combined NASA, Department of Defense (DOD, including SDIO), and Department of Energy space power programs, independent of the extent to which SDI itself is funded.**

1
Introduction

This study focuses on how to satisfy space-based power requirements of the Strategic Defense Initiative Organization (SDIO). (Appendix A is a glossary of abbreviations used in this report.) The initial charge for this study included these tasks:

- Evaluate the planning for the development of advanced space-based high-power technologies to determine the best combination of technology options that should be pursued.
- Critique current Strategic Defense Initiative (SDI) power development plans and objectives.
- Identify an alternate power program plan that would meet SDI requirements for space-based power.
- Identify technology development approaches that could lead to enabling power system capabilities for future space-based defensive systems.

To examine the relevant but less demanding power needs of other U.S. space missions, the scope of the study was subsequently broadened to include consideration of military space power requirements other than those of SDIO, and of potential civil space power requirements, especially those of NASA, where power will be needed for earth-orbital, interplanetary, and lunar-surface missions.

In forming the Committee on Advanced Space Based High Power

Technologies to conduct this study, the areas of expertise sought included nuclear, chemical, and solar energy conversion systems; space environment; materials science; thermal management; power conditioning and control systems; rotating machinery; pulsed-power generation; and system safety. Biographical sketches of the members of this committee are contained in Appendix B.

The committee has had access to classified briefings and publications to provide it with adequate insights regarding activities relevant to the SDI power program and to this study. However, there is no classified information in this report.

The study chronology is summarized in Appendix C. The committee devoted particular attention to briefings and information supplied by the Independent Evaluation Group (IEG), a panel of experts under the leadership of R. Joseph Sovie, NASA Lewis Research Center. The IEG was established by the SDIO Power Program Office to provide it with analyses and counsel regarding its programmatic activities.

The committee held a meeting in Albuquerque designed to gather up-to-date information from the IEG's Field Support Team centered at Sandia National Laboratories. That team provides the IEG with technical analyses of the ongoing power system architecture studies, and includes personnel from NASA/Lewis Research Center, Sandia National Laboratories, and the U.S. Air Force Space Technology Center.

At the outset of the study, the SDIO Power Program Office made available to the committee a number of reports (listed in the References) summarizing the results of contractor-performed studies of a variety of power-generating systems. Because of the ongoing nature of these studies, and recognizing certain shifting SDIO priorities, the committee was briefed on current SDIO thinking in October 1987 and January 1988. Military needs for space-based power for non-SDI applications were described by Air Force and Army representatives. The committee also obtained presentations relevant to projected NASA space power needs from Raymond S. Colladay, then NASA Associate Administrator for Aeronautics and Space Technology, and J. Stuart Fordyce, Director of Aerospace Technology at NASA Lewis Research Center.

The committee sought to keep abreast of relevant activities and progress in a variety of technical fields, through the contacts, expertise, and efforts of its members. For example, the committee examined implications to space power of recent research on high-temperature superconducting materials, of progress in defining the NASA Space Station, and of the growing interest in returning to the lunar surface. The committee also took the initiative of reviewing the status and potential of a system that would provide power to space vehicles by beaming electromagnetic energy to orbit from the earth's surface.

A forerunner to this study, with emphasis on space nuclear power, was conducted by the Committee on Advanced Nuclear Systems of the National Research Council, under the chairmanship of John M. Deutch. That study led to the report "Advanced Nuclear Systems for Portable Power in Space" (National Research Council, 1983).

In the following chapters of this report, the committee discusses much of the information it acquired, with emphasis on those elements it believes provide a basis for its findings, conclusions, and recommendations. Chapter 2 summarizes the broad space power requirements of SDI, other military missions, and civil missions. It also examines approaches to selecting space power technologies to satisfy SDI requirements. Chapter 3 examines available space power system options and some important safety and environmental constraints influencing their selection. Chapter 4 covers the kinds of technological advances needed to meet SDI requirements, and Chapter 5 suggests approaches toward achieving such advances. Chapter 6 examines the current SDI space power R&D program and provides suggestions on how to facilitate its achieving SDI requirements.

2
Space Power Requirements and Selection Criteria

OVERVIEW OF SPACE-BASED POWER REQUIREMENTS

Power system requirements for U.S. space applications can be considered based on the needs of three categories of users: SDI systems, military systems other than SDI, and civil missions. At this time, the requirements of these user categories have only been broadly defined. Summaries follow of these three broad sets of requirements and their commonalities.

SDI Power Requirements for Housekeeping, Alert, and Burst Modes

New U.S. requirements for space-based power imposed by the SDI program greatly exceed space power system capabilities available from past civilian and military experience in spacecraft. In addition, SDI systems and their power subsystems must be survivable in wartime, as discussed later in this chapter, and in the face of possible peacetime attrition attacks while traversing Soviet territory.

SDI applications require electrical power for directed energy weapon (DEW) and kinetic energy weapon (KEW) systems; for surveillance, acquisition, tracking, and kill-assessment (SATKA) systems; and for command, communications, and control (C^3) systems. SDI's requirements also include the integrated and mission-coordinated power-conditioning technologies that are needed to convert the source power into the form required by the specific load being driven. One or more power systems are needed by each SDI spacecraft, potentially at multikilovolt levels, to satisfy power needs categorized by three modes of operation:

- *Housekeeping mode.* Electric power ranging from several kilowatts to tens of kilowatts (or hundreds of kilowatts, if refrigeration of cryogens is necessary) is needed continuously for long periods of time—durations of up to about 10 years—for baseload operation of the space platform, including communication, station-keeping, and surveillance systems. A typical household consumes energy at the average rate of about 1 kW (i.e., 1 kilowatt-hour per hour).
- *Alert mode.* In the event of a hostile threat, powers from about 100 kW to about 10 MWe might possibly be required. Currently, among the three SDI power modes, the alert-mode requirement is the least clearly defined. The total duration of power needs while in an alert status or during periodic testing might even be a year or more. The power level and duration required for the SDI alert mode appear to depend on a postulated operational cycle that is not easily defined, and may also include power for periodic status-checking. Alert-mode power requirements are likely to be higher than can be accommodated with energy storage at reasonable energy storage system masses. Otherwise, either a power system—probably nuclear—would have to be provided or excessive storage capability would be required. Accordingly, unless considerable effort is made to develop SDI systems that minimize the alert-mode requirement, there may be so many kilowatt-hours of energy storage needed—especially if nonnuclear power subsystems are used—that the prime power subsystem would become a major factor in sizing the orbital platform.
- *Burst mode.* For weaponry and fire control during battle, power needed for the burst mode may extend from tens to hundreds of megawatts (and beyond) for durations of a few hundred to a few thousand seconds, and these power levels must be available quickly on demand. Commercial power plants fall into about this power range.

For the alert and burst modes, the unprecedented high power

TABLE 2-1 Salient Features of the SDI Housekeeping, Alert, and Burst Modes

SDI Mode	Duration	Start-Up	Availability	Power Level
Housekeeping	Many years	Not critical	Continuous	Several to 100s of kWe
Alert	Uncertain (up to 1 year)	Minutes	Sporadic (on demand)	100s to 1,000s of kWe
Burst	Minutes	Seconds	Sporadic (on demand)	10s to 100s of MWe

levels, durations, integrated energies, and time-profiles far exceed any current experience with space power systems. Table 2-1 outlines salient features of these three SDI power modes.

Power requirements need to be considered for the major potential SDI weapons and sensor systems that the SDI program is pursuing. Those systems include the following:

- ground-based free-electron lasers (FELs);
- space-based FELs;
- ground-based excimer lasers;
- neutral-particle beam (NPB) systems;
- charged particle beam (CPB) systems;
- kinetic energy weapon (KEW) systems;
- chemical lasers;
- radars (radio detecting and ranging systems); and
- lidars (light detecting and ranging systems).

The SDI Space Power Architecture System (SPAS) studies (1988) indicate that current ground-based versions of FELs and excimer lasers would require prime power in excess of 1 GWe (perhaps as high as tens of gigawatts) at each site; power needs for space versions of these devices are yet to be determined. Space-based free-electron lasers, charged particle beams, and neutral particle beam systems may require from 50 to about 200 MWe per platform; chemical lasers may require only tens of kilowatts.

As summarized in Table 2-1, space-based weapons platforms will require continuous housekeeping power of tens to hundreds of kilowatts (perhaps into the megawatt regime for advanced radar/sensor platforms), depending on the specific system. Power will be needed for refrigeration, communications, radar, and other continuously operating systems. Most SDI space systems are expected to be deployed at altitudes ranging up to a few thousand kilometers, while SATKA platforms, communication/weapon relay, and other special-purpose platforms may be deployed in a range of low earth to geosynchronous orbits. System life is ultimately intended to exceed 10 years if intermittent servicing is feasible on an as-needed basis.

These power requirements are extremely critical to the design of any orbiting platform; severe mass and cost penalties accompany undue conservatism with respect to power level or duration. Depending on the specific SDI system, power subsystems are estimated in the SPAS studies to make up some 20–50 percent of the total mass of the space platform.

The SPAS studies and related data showed that no present integrated technologies could satisfy these ranges of power requirements. Even for the 1990–1995 period, initial estimates of prime power requirements for electrically energized DEW systems tests at the White Sands Missile Range indicate that extremely high power levels having fast time-ramping capabilities must be provided during the tests. Only highly efficient prime power or power conversion technologies could qualify for space-based versions of such applications.

Although the major challenges in developing power technology for SDI applications are associated with space-borne systems, SDI also has power requirements for high-power, ground-based weapons systems. In addition to sources of prime power for ground-based and space-based SDI systems, new forms of energy storage for delivery of the burst mode in space and on the ground may be needed to meet the simultaneous requirements of power level and running time. Satisfying requirements for "instant-on" operation would necessitate development of new ways to switch both power sources and loads. An experiment likely to be relevant to testing the ability for rapid start-up of both ground-based and space-based power systems is expected to be performed as part of the SDIO superconducting magnetic energy storage (SMES) project now under way.

A space environment poses many problems affecting integration and feasibility that have not been previously encountered in either ground-based or space-based systems. Examples of major concerns include source-to-load power transmission; the close physical proximity of source and load; and the large magnitude of the power being transmitted compared to current power levels (a few kilowatts) being used in space.

Requirements of Military Missions Other than SDI

A number of non-SDI military space missions—for example, those of the U.S. Air Force—will require advanced power sources for applications that include surveillance, tracking, and communications. Such applications (Johnson, 1988) focus on hardened surveillance systems as well as electric propulsion systems for orbital transfer vehicles to position military assets in more favorable high orbits. These applications are provisionally projected (USAF/DOE, 1988) to require perhaps 5 kWe in the near term, up to 40 kWe for the midterm, and up to hundreds of kilowatts in the long term; steady generation of power will be the rule. Such space power requirements are technically satisfiable with nuclear reactor or solar power systems, but the size of the required solar arrays could present problems relating to detectability or maneuverability. Current U.S. activity toward developing a space nuclear reactor system (known as SP-100) is directed toward achieving a nominal 100-kWe power output. Subsequent variants of that design may be possible over the power range from 10 to 1,000 kWe.

Power requirements for military, non-SDI space applications will probably overlap those of civil space missions. Military spacecraft requirements include power both for electric propulsion and for onboard uses. A significant additional requirement that power systems for military applications must satisfy will be their survivability in the presence of a hostile threat.

Survivability considerations include needs for military spacecraft to be maneuverable and to have both the capability of being hardened against enemy weapons and of avoiding detection. These considerations impose certain constraints on candidate space power systems, such as the size of solar arrays and the temperature of radiators needed to reject heat to space.

Power requirements for non-SDI military missions can probably be satisfied with solar dynamic or small nuclear reactor power systems. The choice between using a solar or nuclear system may depend on various factors, especially specific mass (measured in kilograms per kilowatt). Future use of advanced Brayton or Stirling cycles could make the solar dynamics option competitive with the nuclear option at power levels of 60 kWe or greater. On the other hand, using dynamic power cycles in space nuclear reactor systems could make nuclear systems more attractive from the standpoint of reducing their specific masses.

Requirements of Civil Missions

Among its current approved missions, NASA's largest projected near-term power need is for the Space Station. Future missions, such as establishing a lunar base and traveling to Mars, will probably require significantly greater power. Solar power systems will suffice for most NASA requirements in earth orbit, but space nuclear reactor systems will probably be needed for planetary and deep-space missions, as discussed in a survey of such needs by Mankins et al. (1987).

NASA options such as space-based materials processing facilities, located in earth orbit or on the lunar surface (Colladay and Gabris, 1988; Ride, 1987), would have power requirements in the hundreds of kilowatts or greater. The Jet Propulsion Laboratory survey (Mankins et al., 1987) of possible NASA needs for nuclear power sources lists approximately 20 possible missions with power requirements ranging from tens to hundreds of kilowatts.

For Phase 1 of the Space Station's development, 75 kWe of average power* will be available from a system of photovoltaic arrays and storage devices. The total area of the solar cell arrays needed to achieve this average power level exceeds 2,000 m^2. Although the earth's atmosphere is extremely tenuous at the station's orbital altitude, atmospheric drag on this very large area of solar cells would periodically require reboosting of the station itself to maintain its orbital altitude. NASA plans to reboost the station by burning gaseous oxygen and hydrogen, obtained by electrolyzing excess water; thus, reboosting would not require fuel supplies from earth.

Once started, photovoltaic space power systems used for civil applications typically operate reliably at their rated average powers for their entire useful lives. Usually the spacecraft for such missions rotate around the earth and experience day and night during each orbit. Thus the energy input into these power systems will ramp up and down once each orbit, necessitating reliable power conditioning and on-board storage.

For Phase 2 of Space Station's development—to increase the average power generated to 125 kWe (300-kWe peak)—NASA's 1987

*Requiring a peak power input of about 200 kWe.

planning called for adding two solar-dynamic power systems* to the Phase 1 power supply. Because of the greater overall efficiency of a solar-dynamic power system compared to that of a photovoltaic space power system, a solar-dynamic system can produce a unit of power from less collection area than is required for an array of solar cells. The improvement in system efficiency results from advantages in thermal storage versus battery storage and from the increased conversion efficiency of a solar-dynamic power cycle compared to solar cells.

Commonality of Requirements Among Civil and Military Missions

While the most demanding space-based power requirements are those of SDI, some projected civil or NASA applications under discussion could capitalize on the SDI investment. For example, to operate an outpost on the lunar surface, a power plant suitable for the SDI housekeeping mode may suit the utility needs. Such a power-generating capability might also be applicable to providing future communications satellites with the capability for a direct-broadcast mode of operation. The burst-mode capability might be useful for powering a catapult on the lunar surface, a device that conceivably could be a factor in making mining of the lunar surface (Kulcinski and Schmitt, 1987) practical. The alert and burst modes may also be useful for spacecraft propulsion. These potential applications are speculative, pending further study.

Long life and reliability are desirable qualities for all space power systems. In addition, many potential missions that have been studied will have power needs that significantly exceed the capabilities of any previous space power sources. These much higher power outputs will require the development of technologies leading to advanced power system components.

APPROACHES TOWARD SELECTING SPACE POWER TECHNOLOGIES TO MEET SDI REQUIREMENTS

Studies completed to date do not provide a basis for selecting a preferred SDI power system or for ranking preferable systems, but

*A solar-dynamic power system converts solar radiation into high-temperature heat, then uses the heat to drive a thermodynamic power cycle.

they do point to areas of leverage. These are areas of technology, mission requirements, and program emphasis where early, careful attention is likely to be cost-effective in: achieving savings in mass, cost, and/or component development time; improving reliability; and in ultimately establishing feasibility (See Chapter 4).

SDI missions impose electrical power requirements far exceeding the state of the art; in particular, to power weapon systems in the burst mode and to supply high-power-demand sensors during the alert mode. These requirements dictate power systems having capabilities ranging from hundreds of megawatts for hundreds to thousands of seconds to supplying several megawatts for times totaling as much as a year to support system operations under alert conditions. Bimodal operation providing both continuous and burst power capabilities may effectively address the combined mission requirements. The multimegawatt technology task of the SDIO Power Program should address these needs by providing for development of an integrated power technology base that considers both nuclear and nonnuclear multimegawatt power sources and combinations of those sources.

For each concept, the following related electrical power supply subsystems should be considered:

- energy source (a source of heat or voltage);
- heat transfer and rejection (thermal management);
- power conversion;
- energy storage (if needed);
- power conditioning and control;
- power transmission; and
- transient performance.

During this consideration of possible multimegawatt power source concepts, a parallel program of review, analysis, and testing of applicable technologies should be conducted to ensure that feasibility issues associated with the systems concepts can be resolved. In many cases, proposed power system operating designs will result in extremely stringent operating conditions, including high temperatures, high pressures, and corrosive substances. The effects of radiation, micrometeorites, space debris, and microgravity on system operating components and materials must also be considered.

Research issues include demonstrating technological feasibility for such considerations as:

- long-term autonomous operational reliability of high-power systems in both natural and perturbed space environments;
- minimizing system mass and size;
- employing higher temperatures; and
- using lower-mass structural and shield materials.

The technology overview should definitely include the following factors peculiar to nuclear power sources:

- radiation safety;
- reactor fuels;
- neutronics and control;
- shielding; and
- reactor thermal hydraulics.

In addition, relevant technologies should be included that affect all power sources, such as the following:

- materials;
- thermal management;
- energy storage; and
- energy conversion and storage.

Several critical issue areas in satisfying SDI space power requirements are discussed below without attempting to rank them by their relative importance; all of them may be vital.

Critical Issue Areas

Figure 2-1 (based on the SPAS studies), which does not include weapons coolant mass, illustrates the sensitivity of the specific mass (measured in kilograms per kilowatt) of space power systems to two critical assumptions: open versus closed cycles* and operating time. For example, an open-cycle space power system might combine hydrogen and oxygen, then discharge the resulting water to space. There may be effluents from the spacecraft even if a closed-cycle power system is used, since military weapons in the spacecraft payload may require a coolant such as liquid hydrogen, which can then be made available to the space power plant as fuel before being discharged. It is tempting to conclude from Figure 2-1 that open cycles

---

*Power systems are classified into those that utilize an open cycle or a closed cycle, according to whether they discharge or recirculate a working fluid, respectively.

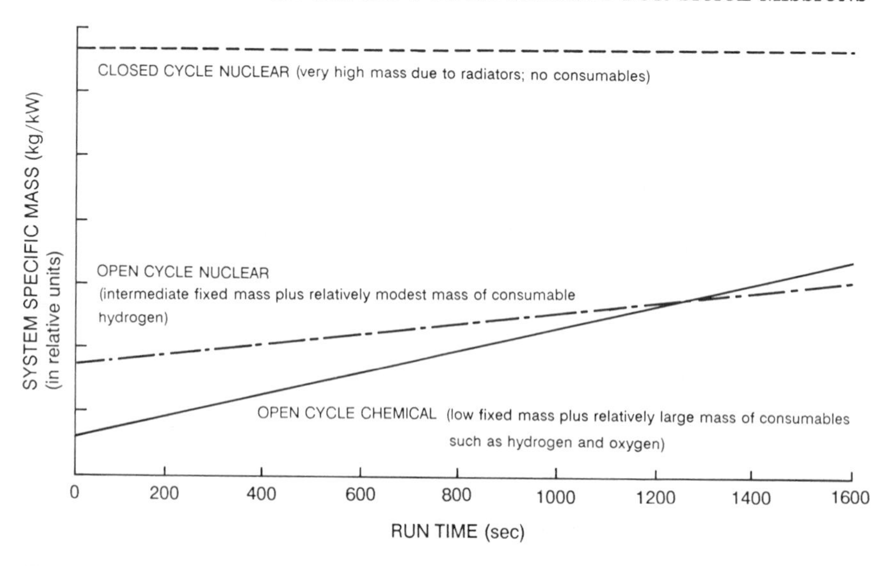

FIGURE 2-1 Sensitivity of system-specific mass to choice of power system and to duration of power use. (Masses of consumables required for weapon cooling are not included in the calculations.) SOURCE: Space Power Architecture System studies, Sandia National Laboratories, and NASA.

must be made to work—at least for the burst power mode—or the entire SDI concept may be very severely penalized.

For the higher-power burst mode, the need for hundreds of megawatts—rising from zero or near zero to full power in a few seconds—for a comparatively short period imposes drastic demands on the system designer. It is presently unclear what penalties are exacted as the price for achieving rapid (i.e., several seconds) start-up times; increasing these times by a factor of two or more could reduce system mass and complexity. With the exception of turbines, all of the power system components being proposed for space applications are massive. Those components include magnetohydrodynamics (MHD) channels, radiators, fuel cells, and power-conditioning equipment.

To minimize total system mass aboard a spacecraft, open-cycle systems that exhaust their working fluid into space are an attractive option. To be successful, this option must include a means to cope with possible adverse effects of releasing effluents. Both the effluent question and the rapid start-up consideration are issues that suggest the necessity of careful review of mission requirements and of the desirability for emphasis on weapons concepts that require only modest power levels.

System Considerations

Satisfying SDI space power requirements necessitates a system approach. Descriptions of nuclear and nonnuclear options sometimes overlap in the following discussion, and some hybrid systems concepts must be covered under the description of nuclear options, particularly in their relation to driving the development of electrical component technology.

Three Space Power Architecture System (SPAS) studies (1988) were performed for the SDIO Power Program Office. The studies were designed to consider and analyze system factors in SDI architecture that define space power requirements in scale, in state of technology, in time, in transient capability, and in reliability. The SPAS studies were also intended to provide guidance for making step improvements in system performance through integrated technology development.

The SPAS studies addressed individual space power system options. However, the spacecraft power supply needed to satisfy requirements for the three SDI power modes (housekeeping, alert, and burst modes) may well not be a single system, but rather an integrated set of generating and power-conditioning *systems* that optimize total-life performance and reliability. Such approaches may well enable power systems that would otherwise remain impractical.

Life-cycle costs will likely be a major factor in the selection of all weapons systems. However, cost considerations were not included in the SPAS studies.

Qualification of Power-Conditioning Subsystems and Components

To qualify for meeting SDI requirements, there must be an adequate experience base for power-conditioning subsystems and components. The committee believes that projections of component performance must be developed based on:

- an experimental data base for component performance;
- analytical models that are anchored to the data base and that permit future capability to be projected; and

- basic research that anchors the model to fundamental processes.

The data base, models, and fundamental understanding must provide technology projections (at required reliability) to assess where additional funding is required. Considerable experience has been obtained for power-conditioning components used with pulses, including elements such as thyratrons, diodes, capacitors, inductors, and transformers. In contrast, less experience has been obtained in the areas of power conversion and conditioning components, such as high-voltage inverters, alternators, generators, and compulsators.

Experience suggests that there will be an optimum load-driven power-module size, as found by designers of accelerators, radar, and electrical power systems many years ago. An analysis of this nature can be applied to SDI power needs and matched to the megawatt average power class of most of the conversion devices. The optimum module size will depend on conversion efficiency, thermal management, power flow, and voltage levels, and may be in the same power range already experienced in the very-high-power radar and fusion fields; namely, between 1 and 10 MWe.

Keeping components and subsystems small and modular also enables local control of faults and minimizes development time. Local fault-control approaches are likely to be required for these very-high-power systems, since only a short period of delay in clearing a fault will destroy the power system.

Influence of SDI Survivability and Vulnerability Criteria

A fundamental SDI requirement is that a space power system, like any SDI system, must be technically effective, cost-effective, and survivable in the face of natural or hostile threats. These three cornerstone requirements are known as the Nitze criteria. Attaining SDI goals—of crisis stability and arms-race stability—would require satisfying these criteria before system deployment. The difficulty of simultaneously satisfying all three Nitze criteria can lead to frustration, which can motivate finding a creative solution for providing power or developing weapons that require less electrical energy.

Assuring a high probability of survival of each system element can be quite costly, both in economic and launch-weight terms, hence survivability is best treated as a system issue. Accordingly, the system designer must balance capabilities for maneuvering, shooting back in defense, decoying, and hardening to provide the required system survivability at minimum cost and launch weight. System trade-offs must include consideration of uncertain parameters such as the threat and technical effectiveness of postulated weapons. These and many other uncertainties tend to lead one to delay trying to satisfy the survivability requirement of system components using advanced technology until that requirement has been more definitively specified and validated.

Pending such a definitive specification, SDIO—in conjunction with the IEG and other advisory groups—has adopted (SDIO, 1986) the interim approach of formulating a list of general guidelines for survivability. Values are listed in that publication for maneuvering, hardness against x rays, and so on, and are probably satisfactory as interim survivability guidelines except for platforms in low earth orbit. Although minimal, these survivability requirements are nevertheless very stressing, hence applying them in the meantime to evaluate the relative survivabilities of otherwise comparable candidate technologies may promote some progress.

There are differences in viewpoint as to how early in the system development cycle one should consider survivability requirements and when there should be an insistence on high levels of survivability. If the system design evolves without survivability in mind, compromises to benefit one criterion may jeopardize survivability. For example, having hydraulically interconnected parallel paths to the many panels of a heat exchanger improves reliability but makes the system fatally vulnerable to a single hit. Incorporating survivability considerations from the outset might lead to thermally interconnected—but hydraulically separated—coolant loops for both reliability and survivability.

Some technologists prefer to emphasize the survivability criterion from the outset, while others recommend postponing survivability issues. The first group argues that applying the criterion early would avoid pursuing inappropriate technologies and would also stimulate new ideas that might be able to satisfy all of the criteria simultaneously. The other group recommends allowing initial research and development on candidate concepts to proceed unfettered by survivability constraints, in order to avoid the risk of prematurely precluding any promising but undeveloped options.

Many technologists are in the first group, while many system architects, such as those who performed the SPAS studies, are in the second group. In the SPAS studies, none of the power systems the contractors examined were hardened prior to estimating masses.

While the committee sees merit on both sides of this issue, it reached no final conclusion as to when in the development cycle survivability should be emphasized. It was apparent that the subject could be handled only at the SDI system level, not solely at the power system or component level.

Rigidly applying survivability concerns to space power systems now would mean there could be comparative studies of only hardened power systems, there could be no development of the largest space-based consumers of electrical power (those requiring more than about 100 MWe), and weapons requiring minimal energy per kill would be favored. Such restrictive actions at this time are unwarranted.

Accordingly, Recommendation 7 below would elevate the concern for survivability to acting as a stimulus to innovation in the development process, and at this stage of exploratory development the committee regards that stimulus as sufficient.

Findings, Conclusions, and Recommendation

Based on the preceding discussion, the committee developed the following findings, conclusions, and recommendation:

Finding 1: Of the three significantly different SDI modes of operation (housekeeping, alert, and burst mode), requirements for the alert mode are inadequately defined, yet they appear to be a major design determinant. For that mode, the unprecedented high power levels, durations, and unusual time-profiles—as well as the associated voltages and currents—that are envisioned will usually make extrapolation from previous experience quite risky and unreliable. A possible exception is in the area of turbine technology, where an adequate range of power levels has been validated for terrestrial applications, although not for spaceflight. Proposed space power systems will need to be space-qualified for long-term unattended use.

Finding 5: Among the power systems that are candidates for SDI applications, the least massive, autonomous self-contained space power systems currently being considered entail tolerance of substantial amounts of effluent during system operation. The feasibility of satisfactorily operating spacecraft sensors, weapons, and power systems in the presence of effluents is still unresolved.

Conclusion 1: Multimegawatt space power sources (at levels of tens to hundreds of megawatts and beyond) will be a necessity if the

SDI program is to deploy electrically energized weapons systems for ballistic missile defense.

Conclusion 4: The rate of rise ("ramp-rate") from zero to full burst-mode power level appears to be a critical requirement. It is not apparent to the committee what relationships exist among elapsed time for power buildup and system complexity, mass, cost, and reliability.

Conclusion 8: Survivability and vulnerability concerns for SDI space power systems have not yet been adequately addressed in presently available studies relevant to SDI space power needs.

Recommendation 7: After adequate evaluation of potential threats, further analyze the subject of vulnerability and survivability, mainly at the overall system level. Data resulting from implementing Recommendation 1 would be appropriate for this analysis. Pending such analysis, candidate power systems should be screened for their potential to satisfy interim SDIO survivability requirements, reserving judgment as to when or whether those requirements should constrain technology development. Convey the screening results to the advocates of those candidate power systems, to stimulate their finding ways to enhance survivability as they develop the technology.

3
Space Power System Options and Selection Constraints

SUMMARY OF AVAILABLE SPACE POWER SYSTEM OPTIONS

The classes of space power systems that are capable of meeting the special prime power and power conditioning requirements of SDI space system architectures are based on three approaches: nonnuclear space power systems, nuclear space power systems, and ground-based power systems for beaming power to space. The nonnuclear options refer to solar photovoltaic, solar-dynamic, and chemical (including magnetohydrodynamic [MHD]) systems. Nuclear options include radioisotope thermoelectric generators (RTGs), dynamic isotope power sources (DIPS), and nuclear reactor systems (see Figures 3-1, 3-2, and 3-3). These power sources can be utilized in "closed" or "open" thermodynamic systems.

Closed and open thermodynamic systems are defined as follows: A closed-cycle system is one in which a working fluid is heated, does work, rejects heat, and is recycled. Closed thermodynamic cycles have various designs. Three important varieties are known as Brayton, Rankine, and Stirling cycles. A Brayton cycle is a conventional closed-cycle employing a gas turbine, in which the working fluid is gaseous throughout the power-generating loop; a Rankine cycle is like a conventional steam cycle, in which the vapor is liquefied in a condenser; and a Stirling cycle is a closed-cycle reciprocating engine

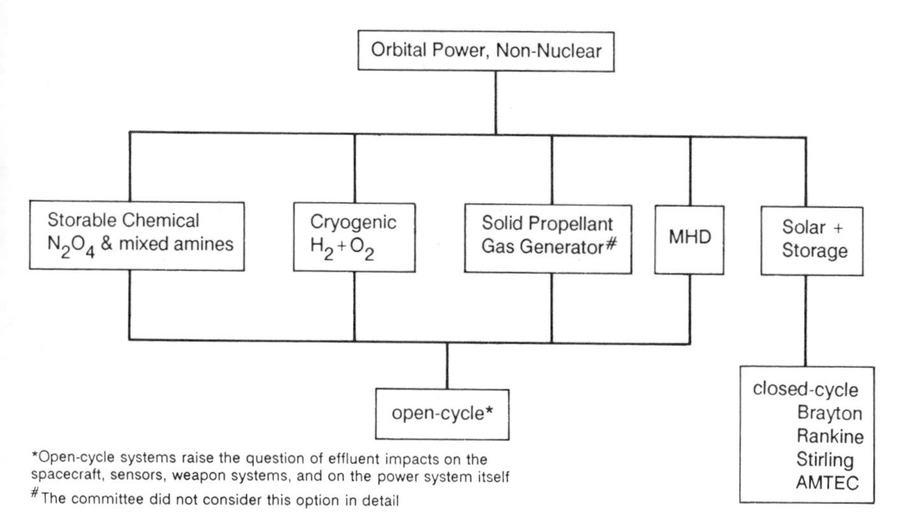

FIGURE 3-1 **A nonnuclear orbital power source.**

whose working fluid is a high-pressure gas, either helium or hydrogen. The alkali-metal thermoelectric converter (AMTEC) cycle is similar thermodynamically to the Rankine cycle turbogenerator system. AMTEC utilizes high-pressure sodium vapor supplied to one side of a solid electrolyte of beta alumina, causing low-pressure sodium vapor to be removed from the other side. Sodium ions transported across this electrolytic membrane produce a voltage difference, which drives electrons through a useful load, whereupon they reunite with the sodium ions to complete the circuit. In an open-cycle system, a working fluid is heated, does work, and is discharged, carrying waste heat with it. An open-cycle system is also a "single-pass" system, since the working fluid is used only once. A variation of the open cycle is the use of chemical reactants—following an exothermic chemical reaction such as combustion—to produce a pressurized vapor and liberate heat. In order to minimize the impact of the resulting effluents on the overall spacecraft system, the reaction products can be separated and, conceptually, some or all of them could be retained, but, in practice, the retention option may prove to be difficult to achieve or totally unrealistic.

Another category of open, or single-pass, systems is one that has no thermodynamic working fluid, or prime mover. Examples are batteries and fuel cells. Usually—but not always—such devices store their effluents.

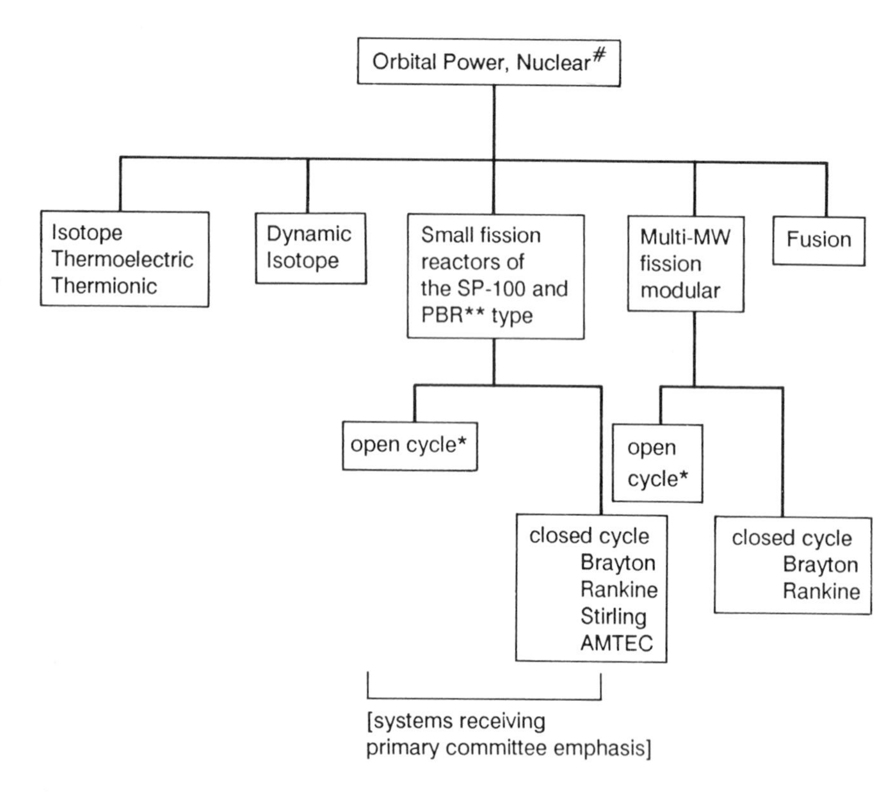

#All nuclear systems introduce complex safety requirements, (see discussion later in this chapter)

*Open-cycle systems raise the question of effluent impacts on sensors, on weapon systems, and on the power system itself

**PBR refers to Pebble bed Reactors and Particle bed Reactors, which are distinguishable from each other by the size of their fuel elements

FIGURE 3-2 A nuclear orbital power source.

Although open-cycle space power systems tend to be less massive than closed-cycle systems for operating periods of less than about an hour, they present the problem of spacecraft toleration of their effluents.

The major space power options available for each SDI power mode are presented in Table 3-1 according to whether or not an effluent is produced.

The committee reviewed SDIO briefing documents that summarized results emerging from SPAS studies (1988) that were being simultaneously conducted by three SDIO contractors while this study

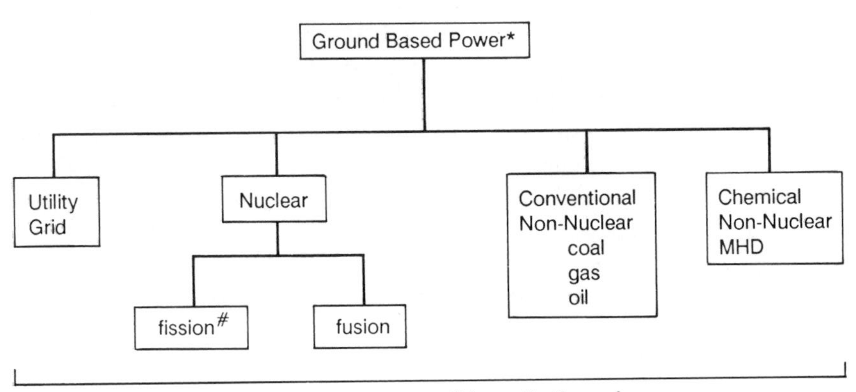

*Requires microwave beam to orbit

#An added benefit might ensue, since this military requirement could--as a spinoff--lead to a true second-generation, fail-safe reactor for civil applications

FIGURE 3-3 Ground-based power.

TABLE 3-1 Space Power Options for Each SDI Power Mode

	Operational Option	
Power Mode	No Effluent	Effluent
Housekeeping	Solar, RTG, DIPS, nuclear reactor, ground-based source	None
Alert	Solar-dynamic, nuclear reactor	None
Burst	Nuclear reactor	Chemical, nuclear reactor

NOTE: DIPS = dynamic isotope power sources; RTG = radioisotope thermoelectric generator.

was in progress. Unfortunately, during the course of this study the SPAS studies did not become available in published form. Information on the SPAS studies was supplied to the committee by the SDIO Power Program Office, by its Independent Evaluation Group (IEG), and by the IEG Technical Support Team. The committee also read the reports of other relevant SDIO-sponsored studies. Although the committee did not review all power system concepts originally considered by SDIO, it did examine several power systems not treated in the IEG summaries.

Figures 3-1, 3-2, and 3-3 are summaries of the power systems reviewed by the committee: (1) nonnuclear power generated in orbit, (2) nuclear power generated in orbit, and (3) power generated on the ground and beamed into orbit electromagnetically. Details on each of these options follow.

Various combinations of prime power generation and storage are possible, but at present it is unclear how to make optimum use of storage—whether by fuel cells, batteries, electrolysis, thermal media, flywheels, superconducting magnetic energy storage (SMES), or by using a combination of those techniques. A power technology research and development program should continue to establish technical feasibility, develop the technologies and resources, and conduct proof-of-principle testing. Applying terrestrial turbomachinery technology to high-power systems for space applications may prove viable but could pose technical obstacles. Areas of uncertainty that will need attention include mass reduction, providing capability for rapid start-up (especially nuclear reactors); operational lifetime, if high-temperature operation is contemplated; embrittlement of turbine blades and other components; in-orbit maintainability; and integration of the power system with its loads and with the space platform.

A major concern for space power generation at the multimegawatt level is thermal management; in particular, the problem of rejecting heat from the power cycle. If a nuclear reactor is used in a multimegawatt space power system, then—unlike a low-power system such as SP-100—the mass of the radiators, rather than the mass of the reactor and its shield, is the dominant component of the mass of the overall power system. Although chemical energy heat sources appear attractive because they offer rapid response and reasonable mass for limited duration, they emit effluents that may have unacceptable impacts. Another potentially attractive option, fusion reactors, will be unavailable—even for terrestrial applications—until well into the next century.

Closed-cycle power conversion systems generate practically no effluents (although attitude control and weapon cooling may produce some additional effluents) and require much lower storage and replenishment of expendables than do open cycles. On the other hand, compared to open-cycle systems operating for short durations or cases in which their working fluid comes from weapon cooling, closed-cycle systems tend to be more massive and require large radiators to reject heat.

The choice of optimum temperature levels for power conversion depends on the selection of conversion cycle and materials, but can be summarized as follows: High heat-addition temperatures* aid thermodynamic performance, but pose materials problems. Low heat-rejection temperatures improve thermodynamic performance, but result in large, massive radiators, thus posing vulnerability and maneuverability problems.

Some low-mass concepts for heat rejection are worthy of consideration for heat rejection at low temperatures (i.e., below 1000°K). These include the liquid droplet and liquid sheet radiator concepts, and various moving belt (liquid, solid, and hybrid) radiators. All of these approaches are in the advanced conceptual stages of development, and none of them have been adequately tested. Questions regarding maneuverability issues, particularly for the belt radiators, and contamination caused by escaping fluids have not been addressed. It is anticipated that such systems, if successful, would not be available before the year 2000. Nonetheless, the heat rejection issue is sufficiently critical that such advanced concepts merit consideration for future SDI systems.

A closed-cycle system typically employs a Brayton or Rankine closed loop and a turboalternator for power generation. Although activation of a chemical heat source can be rapid, it may be more difficult for a fission reactor to reach full power quickly. Consequently, substantial housekeeping power may be required to maintain the power conversion cycle components in a warmed-up condition ready for rapid start-up. Bimodal operation of a nuclear power plant to supply power for both the alert and burst modes would substantially

---

*Currently, temperatures for metallic parts of terrestrial gas turbines can go to 1100°C (1373°K). Refractory metals, if oxygen is completely absent, offer reasonable hope of attaining 1500°K–1600°K in space (Klopp et al., 1980).

alleviate its start-up problems. The reactor's increased power output in going from alert-mode operation to burst mode could then be achieved with only modest change in its operating temperature, in contrast to the significant changes in temperature and heat output that would occur if the reactor must be activated just prior to the burst mode.

Nonnuclear Power for Orbital Use

Nonnuclear (largely chemical and solar) power-generating systems (Figure 3-3) offer the attractive feature of avoiding certain safety concerns associated with nuclear systems. Furthermore, the gross mass of open-cycle chemical systems is less than that of open-cycle nuclear systems for the short operating times typical of SDI burst-mode power requirements. In these open-cycle systems, a major effluent is ***hydrogen***, which is typically used for weapon cooling. Chemical power generation systems may also emit water or other reaction products, although these products could conceivably be condensed and stored on board. Presently it is unclear whether any or all of these effluents can be tolerated by SDI systems; however, the effluent question is clearly an important consideration in choosing between effluent-emitting and closed-cycle nonnuclear power systems for space applications.

The several levels of power required by the housekeeping, alert, and burst modes are significant discriminators among candidate nonnuclear systems.

Photovoltaic Space Power Systems

The most commonly used long-lived space power systems are based on the photovoltaic conversion of solar radiation into electric power. The largest such power plant usefully applied in a space mission was aboard Skylab. For that spacecraft, 3.8 kWe of solar power was installed for the Workshop and 3.7 kWe to operate the Apollo Telescope Mount.

There are still substantial problems to be solved. Means for erecting large arrays (about 25,000 m^2/MWe) have yet to evolve, and the structural dynamics of these low-mass and generally flexible arrays remain to be developed. In addition, when high-enough voltages are generated, some interaction will occur between the solar array and the space plasma, resulting in arcing or power losses.

Since arcing can damage the array, electrical insulation in the space environment is a major issue. The arrays cause orbital drag, requiring make-up thrust to compensate. Any magnetic fields generated interact with the earth's magnetic field, producing torques.

Photovoltaic arrays presently used in space are sensitive to hostile action because of their large size, their low mass per unit area, and exposure of their semiconducting cells to the threat. In contrast, in concentrator arrays now being studied, the metallic mirror and supporting structure provide a substantially smaller target and some protection to the semiconducting cells. Concentrator arrays have a very narrow cone of optical vulnerability, centered on the direction of the incoming solar radiation. If the hostile threat is a beam, that orientation is difficult for the beam to achieve.

Solar-Dynamic Power

Solar-dynamic power generation is being considered for the Space Station as a means of reducing the area required for collection of solar radiation compared to the area that would be necessary if all the power were supplied by photovoltaic arrays.

To date, the largest solar parabolic collector built for space power applications was a mirror 6 m (20 ft) in diameter (English, 1978). This solar collector and its heat receiver also require fairly accurate orientation toward the sun, an acceptable pointing error being perhaps 1 to 3 arc-minutes.

Rankine, Brayton, and Stirling power conversion cycles have been proposed for use with solar energy sources. The Stirling cycle employs a reciprocating engine—for which a firm long-lifetime technology base is not yet available—and is attractive primarily because of its high cycle efficiency at moderate temperature. The Rankine and Brayton cycles utilize a turbine driving an alternator. By employing fluid-film, gas-supported, or magnetic bearings, turboalternator wear mechanisms can be avoided, hence long lifetimes appear to be attainable.

Substantial development work on both the closed mixed-gas Brayton cycle and on the organic Rankine cycle engine has been done over the past 20 years, first as candidates for 1- to 10-kWe, isotope-fueled power systems, and more recently as contenders to supply power for the Space Station. Both Brayton cycle and Rankine cycle power conversion systems use large multistage turbines to drive electric generators.

In summary, for generating solar-electric power in low earth orbit, solar-dynamic power plants have the potential to produce several times the power output of solar photovoltaic arrays having the same collecting area. These power plants are constructed almost entirely of metallic materials, and their semiconducting components (chiefly for power conditioning and control) are small, and thus more easily shielded from being damaged by charged particles in space or by man-made radiation. Solar-dynamic power systems to supply from 50 to 300 kWe are being considered for Phase 2 of the Space Station.

Chemical Space Power Systems

Chemical reactants can be stored aboard spacecraft for power generation as well as propulsion. These reactants can be used to power open-cycle power systems, as summarized in Figure 3-2. Considerable technology relevant to this application is available from the extensive technical experience derived from using stored propellants aboard the Titan rocket and aboard spacecraft used during the Apollo program. Nitrogen tetroxide (N_2O_4) and mixed amines are quite easily stored and require no separate ignition system. The associated metals and synthetic sealing materials have been amply demonstrated.

The principal unknown in using chemical reactants to produce space power is the tolerability of the spacecraft systems to the impacts of any chemical effluents that are released. A basic shortcoming in using chemical reactants is that their mass becomes prohibitive for durations beyond about 1,000 s (see Figure 2-1). Cooling of the weapon system would require a separate liquid hydrogen supply, and would also produce effluent, but release of hydrogen may well be tolerable. Space experiments could help resolve the relative tolerability of hydrogen compared to other effluents, such as water (see Recommendation 2).

Space power systems using stored cryogens such as liquid oxygen and liquid hydrogen provide an attractive source of energy for an open-cycle space power system. The Apollo and Space Shuttle programs provide a well-developed background for applying cryogens to propulsion. The turbine-driven fuel pump for the Shuttle's main engine represents a record achievement in horsepower per unit mass. Insulation for cryo-tankage is well understood. The liquid hydrogen supply could be provided by the weapon-cooling system.

The principal penalty for a cryogenic system is the requirement for active refrigeration. There may be some trade-off between system pressure, loss due to boil-off, refrigeration system mass, and insulation mass. Figure 4-4 is representative of such a system.

Both storable and cryogenic space power systems are feasible. The choice between them should be based not merely on comparison of the power subsystems, but on the basis of comparing all-up spacecraft system designs (see Recommendation 1).

Magnetohydrodynamic Space Power Systems

Magnetohydrodynamic (MHD) space power systems, a variety of chemical power systems, are still in the research phase. The basic principles of operation of an MHD electrical power generation system are conceptually simple, although practical systems are difficult to realize. In an MHD system, electricity is generated directly by causing a conducting fluid to flow across magnetic field lines. Such a system operates similarly to a Faraday-disk machine, except that a conducting gas is substituted for a metallic conductor, and linear motion through a channel is utilized instead of rotational motion. Because of this substitution, an MHD generator may be less massive than a conventional generator, hence MHD generators have some prospect for reducing the mass of space power systems, especially for burst-power applications requiring peak powers measured in multimegawatts.

In practice, introducing MHD technology poses several practical problems in addition to its extremely high operating temperatures and the need to obtain adequate electrical conductivity. One category of problem relates to achieving satisfactory behavior of the fluid flow in the MHD channel during the conversion process. Another problem is the management of system effluents emerging from the channel.

Mitigation of the first problem requires attaining a highly ionized, high-velocity gas stream having adequate uniformity. The gas flowing through the MHD channel consists of a mixture of the hot combustion products of an exothermic reaction—which provide fuel—seeded with an alkali metal (e.g., potassium) to improve electrical conductivity when ionized. Small nonuniformities of gas density and/or ionization concentration (conductivity) can result in major flow instabilities, and the excess heating in these regions causes acoustic disturbances and flow disruptions.

The effluent problem requires finding channel geometries that maximize uniformity of flow and minimize excess heating—and the resultant acoustic disturbances—in the conversion and exhaust regions of the channel. The high-MHD generator-exhaust temperatures (about 2500°K) pose difficult materials problems and, if the escaping ionized gas is discharged to space, the glow emitted by the recombination of its ions and electrons would be visible to an enemy.

The total mass flow rate of effluent for a 200-MWe MHD space power system was estimated in the General Electric SPAS study (1988). This mass flow was 24.3 kg/s, which was considerably less than the 39.6 kg/s emitted by a conventional 200-MWe hydrogen/oxygen chemical power system.

The SDIO Power Program Office has recently funded a feasibility study to evaluate the applicability of MHD to SDI requirements. The first phase of that project is for independent feasibility assessments of two candidate concepts for a multimegawatt space power system. The second phase is to assess innovative approaches to develop such systems. These studies will address problem areas such as uniformity of the ionized gas in the MHD conversion channel, channel erosion, and dealing with substantial quantities of metallically seeded ionized effluents. MHD space power systems could degrade spacecraft stability or perturb orbits.

Based on these considerations, the committee considers that the state of the art in MHD technology may eventually warrant demonstration in space. However, until MHD systems that might be developed for SDI are projected to be capable of modifying or trapping such effluents, it is the sense of the committee that further MHD development for SDI—beyond the conceptual studies and scaling validations presently contemplated—is not warranted.

Nuclear Power for Use in Space

Nuclear power technology can provide the capability to satisfy the power-density and power-level requirements for a variety of civil and military missions in space. The United States has used radioisotope thermoelectric generators (RTGs) in space, but has never employed nuclear reactor power systems for space applications except for a short-term test of the SNAP-10A power system in 1965. In contrast, the USSR has continued to develop and deploy fission reactor systems that have been largely successful, although two unplanned reentries of Soviet nuclear-reactor-powered satellites have occurred, causing adverse public reaction throughout the world.

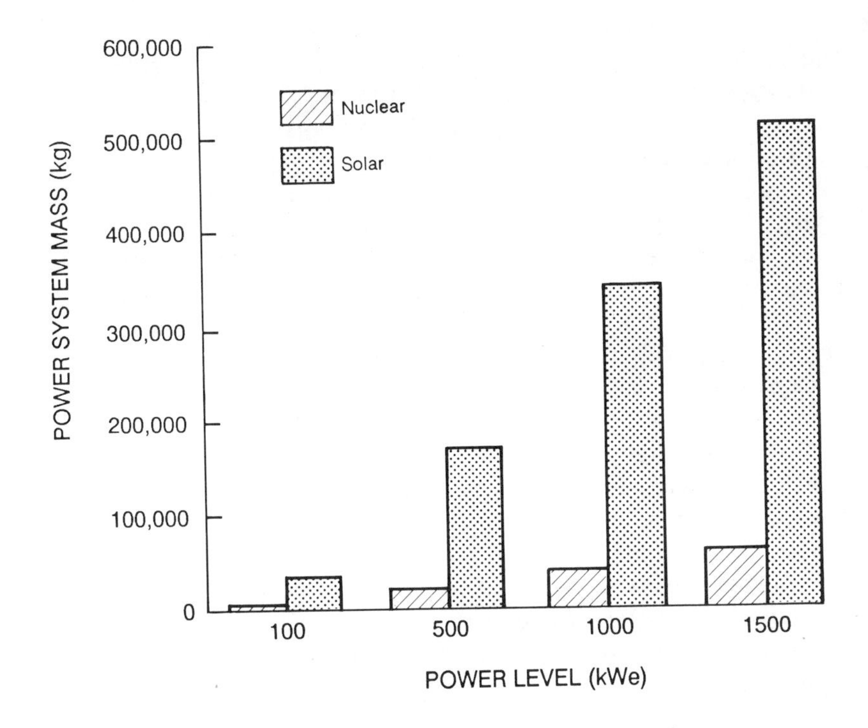

FIGURE 3-4 Mass comparison of lunar power systems. PMAD = power management and distribution. SOURCE: NASA Internal Study, 1988.

The committee believes that there are earth-orbital and lunar surface applications for which nuclear power can be an important and sometimes unique option. For example, Figure 3-4, from a NASA internal study (1988) shows the mass advantage of a nuclear power system over an advanced solar power system for powering a lunar base. The solar power system would have to be much more massive than a nuclear power system in order to provide storage during the 14-day lunar night.

In the following presentations relating to nuclear power in space, the committee examines safety, environmental, and regulatory considerations, then discusses technical aspects of six categories of candidate space nuclear power options. These options offer the possibility of developing compact space nuclear power systems that have favorable specific-mass characteristics and that are suitable for generating energy for long periods. A variety of nuclear reactor system designs have been proposed as candidates to supply steady-state and burst power for both civil and military applications. These candidate systems range from a few kilowatts to several hundred megawatts, and their designs cover a spectrum of technologies. However, with the exception of the SNAP-10A reactor, none of the reactor systems have been tested in space by the United States.

The six general categories of nuclear power sources (Figure 3-2) considered here are as follows: radioisotope thermoelectric generators, dynamic isotope power sources, the SP-100 class nuclear fission reactor, small nuclear fission reactors, and power sources using advanced nuclear processes. A complete nuclear reactor power supply system consists of a nuclear reactor, shield, power conversion system, radiator (or coolant supply for the open-cycle case) and bus power conditioning.

Nuclear Safety, Environmental, and Regulatory Considerations

The committee believes that the responsible approach to developing nuclear power systems for space missions is to make safety of paramount concern, and that it must be designed into candidate nuclear power technologies from the outset. Using a nuclear power source for space power requirements is an important option to preserve, yet poses significant safety risks. For space reactor systems to be regarded as safe, they will need to present an extremely low risk to the biosphere. This is a necessary condition for obtaining public acceptance, but it may not be sufficient.

Safety and institutional acceptability must be considered from both general and operational points of view. Reasonable risk is always difficult to define; however, time can be saved and frustration avoided if concerns associated with nuclear power in space are faced early and openly.

From an operational point of view, the safety of a nuclear power system for use in space must be examined under the various sets of conditions that the system may experience. These include the prelaunch phase (assembly of the reactor, ground testing, assembly of the launch-vehicle cargo); the launch phase (abnormal launch-sequence trajectory, on-pad or suborbital accidents, failure to attain orbit); the on-orbit phase (on-orbital maintenance of spacecraft, power system control anomaly, erroneous signal from the ground, inadvertent spacecraft reentry, and ensuring reactor safety); and the end-of-mission phase (safety of orbit or establishment of safe orbit, ensuring reactor safety).

In a broader sense, overall safety contingencies that must be satisfactorily addressed for a space nuclear system include inadvertent or uncontrolled criticality; protection of the biosphere; protection of occupational workers, astronauts, and the general public against radiation and toxic materials; safeguarding nuclear materials against diversion; disposition at the end of its useful life; compliance with domestic and international law; and achieving public acceptance—that is, the perception that all of the above issues have been handled honestly and reasonably. Certain of these contingencies and some approaches for dealing with them are developed further in the following paragraphs.

Fundamental safety concerns about space nuclear power systems focus on the radiological hazards of prelaunch or launch malfunctions and on unplanned reentry. Prior to and during launch—before it is first brought to criticality—a nuclear reactor has a much lower radiological inventory than later in its life cycle, that is, after fission products have accumulated. From this standpoint, although a nuclear reactor would typically have a much greater power level than an RTG, before initial criticality, a reactor is significantly less radioactive than an RTG. Although it has a low probability of occurrence, the event with the greatest potential adverse impact would be the reentry into the biosphere of a nuclear reactor that has failed to respond to a remotely controlled shut-down instruction.

Inadvertent criticality concerns protection against core compaction that would be caused by launch explosions and high-velocity ground impacts, soil burial, or loss of reactivity control. Additionally, criticality safeguards must encompass water immersion and flooding of a damaged reactor, core configuration and composition changes, and inadvertent removal of neutron absorbers. Levels of exposure to radiation must be kept to acceptable standards for all planned or unplanned activities, such as maintenance, upgrading of components, and accidents. Ensuring end-of-life neutronic shutdown protects against core disruption and release of radioactive material due to power excursion. Protection must be provided for the external heat load due to atmospheric reentry.

A relatively fail-safe approach to deploying nuclear-reactor power systems in earth orbit would be to restrict spacecraft carrying such power systems to so-called "nuclear-safe" orbits in order to reduce deleterious impacts of an unplanned reentry. Such orbits would need to have orbital-decay times that are sufficiently long—300 years or more (Seaton, 1985; Buden, 1981)—to allow ample decay of the radioactive inventory. Because the highly enriched uranium is used for fuel, that inventory is primarily composed of fission products which typically have half-lives ranging from a few seconds to 30 years. A very small percentage of the radioactive inventory consists of longer-lived (transuranic) radionuclides. If a mission requires a nuclear-reactor power system aboard a spacecraft in a lower orbit, the reactor must be boosted to a nuclear-safe orbit after mission completion. The Soviets have launched about 35 nuclear space reactors relying on this approach. On at least two occasions (Johnson, 1986), this approach has failed. The altitude corresponding to a 300-year lifetime depends upon the ballistic coefficient of the system (Buden, 1981); for an SP-100 reactor power system without payload, suitable orbital lifetimes are achieved for altitudes of 800 km or greater.

Environmental, safety, and regulatory considerations need to be dealt with both domestically and internationally. Domestically, the mission agency responsible for the program should file a programmatic environmental impact assessment (EIA) early in the process. From an international standpoint, relevant space and environmental laws are diverse and are often of indirect applicability to military missions.

No single source of international law directly governs the use of space-based nuclear power systems. Instead, six international conventions and a United Nations resolution in some way address nuclear power systems. The Outer Space Treaty (Treaty on Principles Governing the Activities of States in the Exploration and Use of Outer Space) requires a country to consult with treaty members prior to deployment of a system that may contaminate outer space. Should a system malfunction or pose a threat to the outer space environment, consultation is required prior to taking corrective actions. The treaty places the international liability for nuclear contamination on the country that launches a space object, even if the mission is aborted on the launch pad.

The Liability Convention (Convention on International Liability for Damage Caused by Space Objects) expands on the Outer Space Treaty to include damage (loss of life, personal injury, health impairment, or loss or damage to property) on the surface of the earth or to an aircraft in flight caused by a space object. A related convention, the Convention on Third Party Liability in the Field of Nuclear Energy, sets limits for collectable damages.

Radioisotope Thermoelectric Generators

Radioisotope thermoelectric generators have been demonstrated to be useful and reliable power sources to supply a few watts to a few kilowatts of power for space missions. Heat is provided by radioactive decay of plutonium 238 (Pu-238), the half-life of which is 87.7 years. Thermoelectric devices are employed for power conversion. Since Pu-238 undergoes primarily alpha decay, little radiation shielding is required, hence RTGs can be used for manned missions. However, for cost and safety reasons, current RTG designs are mainly suitable only for low-power applications (current units typically produce 0.275 kWe). In addition, RTG conversion efficiency is approximately 5 percent, and power output of RTG systems decreases during service through decay of the plutonium heat source and the degradation of the thermoelectric devices.

Efforts to improve the performance of RTG power systems are focused on increasing their net conversion efficiency. New thermoelectric devices are being investigated that can operate at higher temperatures with improved conversion efficiency.

A primary user issue relating to RTGs is the availability and cost of the isotope Pu-238. Since Pu-238 is essential for a variety of other applications, a significant increase in demand may exceed current production capability. In addition, even if a sufficient supply of Pu-238 were available, its cost (roughly estimated at about $20 million per kWe) would make large-scale use of RTGs impractical. For example, a 1-MWe power source using several thousand RTGs would have a fuel cost alone reaching into the billions of dollars.

The primary RTG safety issue relates to the possibility of a launch accident, in which case contamination could occur in the vicinity of the accident. Although up to now the general public may have been unaware of U.S. rocket-launched RTGs, resistance to future launchings could arise because of concerns about possible launching mishaps.

Dynamic Isotope Power Sources

Because of their higher system efficiency and reduced specific mass compared to RTGs, dynamic isotope power sources are an attractive advanced power option to supply power levels of a kilowatt or greater. They can provide about five times as much power output as an RTG from a given supply of Pu-238, but presently appear to be limited to power outputs of about 5 kWe because of the high cost and limited availability of Pu-238. Dynamic power-conversion cycles that are being considered utilize Brayton cycle, organic Rankine cycle, Stirling cycle, or liquid metal systems. Considerable experience regarding DIPS systems is available from terrestrial investigations, but their long-term, unattended reliability in space must still be demonstrated.

As with RTGs, the principal technical issue to be resolved for DIPS systems—once their feasibility has been demonstrated—will again be the availability of the Pu-238 radioisotope in sufficient quantities and at acceptable cost to make widespread use of DIPS feasible. The primary safety issue confronting DIPS systems, as for RTGs, is likely to be launch safety.

SP-100 Space Nuclear Reactor System

The only U.S. nuclear fission reactor system that has been developed and tested in space, SNAP-10A, consisted of a NaK-cooled 43-kWt reactor with thermoelectric power conversion, and produced 0.56 kWe. SNAP-l0A operated in space for 43 days in 1965, until it failed because its voltage regulator malfunctioned and triggered its automatic shutdown. During the period from 1957 to 1972, federal funding for nuclear space reactors totaled $735 million, but from 1972 to 1982 the United States funded only a modest research activity on space nuclear reactor technology, at a level of about $1 million per year.

The latest program for developing a nominally 100-kWe space nuclear reactor power system was established in 1983. That power system was designated as SP-100, and the program was organized through a tri-agency agreement among DOE, NASA, and DARPA. DOD later formed the Strategic Defense Initiative Organization (SDIO), which replaced DARPA as the DOD representative in the program. The SP-100 program includes development and ground demonstration of a nuclear reactor power system that employs uranium nitride fuel, liquid-lithium coolant, and thermoelectric power conversion. The baseline SP-100 design provides a power output of 100 kWe. Possible future redesign options are contemplated for space power applications over the range extending from 10 kWe up to 1 MWe.

In defining the baseline SP-100 system, it was necessary to select technologies that would both represent advances in the state of the art and have a reasonable likelihood of success. As a result, uranium nitride fuel, clad with PWC-11 (niobium one percent zirconium 0.1 percent carbon), was chosen as the material for operating in the range of approximately 1400°K.

Considerable operational experience with Nb-one percent Zr had been obtained from the previous space nuclear reactor system program. Although higher operating temperatures would be desirable, the creep strength of Nb-one percent Zr decreases at temperatures significantly above 1450°K. Similarly, thermoelectric power conversion was selected because, despite its low conversion efficiency, considerable operating experience with RTGs is available. The major drawbacks of such a system are its net conversion efficiency of about four percent and the resulting need to reject large amounts of heat.

In addition to the baseline SP-100 power system, several alternative materials and subsystems that could substantially improve SP-100 performance are also being investigated. Materials under investigation (Cooper and Horak, 1984) include other clad and structural alloys, such as PWC-11 (Lundberg, 1985), and molybdenum–rhenium alloys. These alloys have significantly greater creep strength than Nb-one percent Zr, but little information is available on their behavior under irradiation. On the power conversion side, two alternatives are being examined: a high-temperature thermionic fuel element and the Stirling engine. The thermionic fuel element would operate at internal temperatures of 1800°K to 2000°K (external temperatures in the 1000°K to 1200°K range), while the Stirling engine, with its significantly higher conversion efficiency, would in the near term operate at 1050°K, permitting the use of special stainless steels or lower-alloy superalloys such as S-590, S-816, or N-155. To make use of the full capability of the SP-100 reactor, the long-range goal of the Stirling power conversion program is to develop a converter operating at 1300°K, which would result in maximum specific power density (watts per kilogram) of the system. The 1300°K Stirling converter would require the use of refactory alloys (such as PWC-11) or composites (e.g., tungsten wire-reinforced Nb-one percent Zr).

The SP-100 program is currently entering into development of a ground engineering system (GES) to demonstrate the nuclear reactor power system during a 90-day test on the ground. A reactor and primary heat transport system will be combined in a full-scale system test. In addition, an out-of-pile, end-to-end demonstration will be performed using 1/12th of the full system, and employing a reactor simulator, the power conversion system, and the heat rejection components. Another objective of the GES tests is to advance fuel-element technology through the use of both advanced ceramic fuels and refractory metal-alloy materials. If these tests are successful, a demonstration of SP-100 in space could follow.

The SP-100 system design incorporates innovative components wherever appropriate, and the SP-100 program is designed to allow opportunities for evolutionary growth. It is recognized, however, that the materials and subsystems employed must be qualified through demonstration at the operating temperatures envisioned and over the desired operating lifetime. System trade-offs—for example, system responses to a reduction in operating temperatures—must also be considered. In that example, system performance would be reduced but the probability of achieving a successful system would increase.

The primary technical goals of SP-100 development are to achieve terrestrial and space demonstrations of a nuclear reactor power system acceptable to planners of space missions that require high powers and high power densities. If the SP-100 engineering design is shown to be feasible, a wide range of civil and military space missions are foreseeable where SP-100 technology could be utilized.

Additional delay in implementing a space nuclear reactor power system program will of course add to the time needed for planning and deployment of an associated space mission. The dilemma is that the development time for such a system still significantly exceeds normal mission-development time, yet a candidate power system must be fully demonstrated to be feasible, reliable, safe, and acceptable to the general public, legislative bodies, and regulatory agencies before any space mission planner can count on utilizing the system as a power source. Hence a primary issue to be addressed by the SP-100 program is that SP-100 satisfy mission safety criteria and requirements. A rigorous review based on the earlier discussion of safety for nuclear power systems is an absolute prerequisite to mission assignment. This situation is recognized in Finding 8.

Smaller Nuclear Space Reactor Systems

A number of missions, both civil and military, have been identified with power requirements in the 5 to 40 kWe range. Studies are being conducted (USAF/DOE, 1988) to determine if an alternative nuclear fission space reactor power system would be preferable to redesigning SP-100 to meet these requirements. Several potential candidate power systems are being considered, including a lower-power version of SP-100.

While there appear to be sufficient mission needs for a long-term energy source in this low-power range, as with SP-100, it will be necessary to demonstrate a small-reactor system before planners would incorporate such a design into upcoming missions. In addition, if an alternative to the SP-100 design were selected for this application, years may be added to the development schedule if new technologies are required. Again, a primary issue of concern will be safety.

Multimegawatt Nuclear Space Reactor System Designs

Various designs have been proposed for nuclear fission space reactor systems capable of producing power in the multimegawatt range. Candidate system concepts are discussed, for example, in the Proceedings of the Symposium on Space Nuclear Power Systems. These Proceedings are available for symposia that are held each January in Albuquerque, N. Mex. Currently, candidate options exist only at the conceptual stage. In addition, multimegawatt systems for SDI will need to satisfy requirements that include providing power for operating in the housekeeping, alert, and burst modes. Both open- and closed-cycle systems are being investigated, as well as several approaches to power conversion; however, closed-cycle systems impose significant mass penalties, while open-cycle systems produce effluents that could impair overall performance (see Findings 1 and 3). The multimegawatt power range can be divided into three regimes: (1) tens of megawatts for durations of 10 years or more, (2) bursts of about 100 MWe, and (3) bursts of hundreds of megawatts. Power sources for the two burst-power regimes probably correspond to closed- and open-cycle systems, respectively.

Tests are being conducted with two candidate technologies: in-core thermionics and the gas-cooled pebble/particle bed core. In the thermionic fuel element demonstration, power conversion is accomplished within the nuclear fuel elements.

Other Advanced Nuclear Systems

New concepts utilizing nuclear processes other than radioactive decay or nuclear fission may play a role in the future. Such systems might, for example, include magnetic fusion reactors. The overriding technical questions will be demonstration of the eventual feasibility of these processes for space power systems. Furthermore, it is not possible to project a date of availability for these advanced systems.

If space power sources are to be developed to provide electricity to power space weapons for defense against ballistic missiles, it seems clear that multimegawatt power sources—either nuclear or nonnuclear—are an absolute necessity, based on current understanding of requirements. With current SDI emphasis on "near-term" deployment, space power systems are being developed only for such functions as surveillance, discrimination, or detection, hence they lack the multimegawatt capabilities required for weapons (Conclusion 7 and Recommendation 3c).

Finding, Conclusion, and Recommendation

Based on the above, the committee arrived at the following finding, conclusion, and recommendation:

Finding 8: The time needed for the development and demonstration of a U.S. space nuclear reactor power system currently exceeds the time required to plan and deploy a space mission dependent upon that power source.

Conclusion 7: A space nuclear reactor power system, once available, could serve a number of applications—for example, in NASA and military missions requiring up to 100 kWe of power or more—in addition to SDI.

Recommendation 3a: Give early, careful consideration to the regulatory, safety, and National Environmental Policy Act requirements for space nuclear power systems from manufacture through launch, orbital service, safe orbit requirements, and disposition.

Ground-Based Power Beamed to Orbit

Power transmission from ground-based power systems for reception and use aboard spacecraft for propulsion or power needs could conceivably become practicable. The ground-based portion of power-beaming systems could employ a combination (summarized in Figure

3-3) of dedicated power plants, energy storage devices, and linkages to the commercial power grid. However, in examining the relative merits of the power-beaming option, trade-offs must be made based on resolving certain problems intrinsic to this option.

One such problem is the vulnerability to hostile threats of fixed ground-based transmitter facilities and—in the case of microwave transmission—of the rectennas used for reception of power at the spacecraft. Another problem is that space platforms will "see" a given transmitter beam during only a small portion of each earth orbit for the low earth orbit (LEO) application. Solving this problem would necessitate using some combination of multiple ground-based transmitters and spacecraft storage of electrical energy. The received electromagnetic energy would be used to charge an on-board energy storage device (e.g., batteries, superconducting magnetic energy storage). The need for energy storage and a rectenna is associated with significant spacecraft mass.

The relative attractiveness of the power-beaming-to-spacecraft option is probably closely linked with the brief time available for line-of-sight transmission (Hoffert et al., 1987) and with the masses of the rectenna and of on-board power storage systems. Existing storage devices, especially batteries, are too massive, although future batteries may qualify. Even if the mass of the storage system combined with the rectenna were attractive, any potential mass benefits of this option must also be balanced against system vulnerability. A rectenna would make the spacecraft difficult to maneuver, and in contrast to the relatively compact SP-100 nuclear option—which includes a significant mass penalty to achieve minimal hardening—is so extensive and fragile that it would be difficult to camouflage or to harden as a target.

The committee considered the limited available information (Brown, 1987; Gregorwich, 1987; Hoffert et al., 1987) regarding the option of beaming power from the ground into orbit using microwaves, and found that this concept has some attractive features. For example, there is the potential for keeping much of the power-generating machinery on the ground, where mass is not critical and where maintenance and refueling are simpler. On the other hand, microwave power-beaming systems do have certain drawbacks: brevity of transmission periods, complex orbital mechanics, the masses of on-board energy storage systems, the vulnerability of large-area rectennas and ground installations, and a possible need for orbiting relay reflectors. The committee regards this option both cautiously and seriously, and believes that further study is warranted to evaluate this system concept.

There has been only a modest evaluation of the various power-beaming-to-spacecraft options that have been suggested. These options include (a) beaming power from earth directly to a weapon; (b) beaming power from earth to a reflector in geosynchronous orbit and thence to the weapon in low earth orbit; and (c) beaming power from earth to a converter in synchronous orbit, thence to a reflector in low earth orbit. A variation on these options would be to beam power downward from a space power source in higher orbit.

Finding and Recommendation

Based on the above, the committee arrived at the following finding and recommendation:

Finding 6: Beaming power from earth to spacecraft by microwaves or lasers (see Recommendation 6) has not been extensively explored as a power or propulsion option.

Recommendation 6: Review again the potential for ground-based power generation (or energy storage) with subsequent electromagnetic transmission to orbit.

Co-Orbiting Power Sources

Power can be delivered to a spacecraft from a detached part of that spacecraft or from a co-orbiting spacecraft by the use of long tethers, rigid booms, or by beaming. The concept of locally beamed or tethered (i.e., via long cables) power transmission from a power source to a weapon "at some distance"—which is taken to mean within a distance on the order of a kilometer or so, appears possible but is probably very complex.

ENVIRONMENTAL CONSTRAINTS INFLUENCING THE SELECTION OF SPACE POWER SYSTEMS

The Natural Space Environment

The natural space environment contains neutral gases, plasmas, radiation (both penetrating particles and solar electromagnetic), magnetic fields, meteoroids, and space debris. Characteristic densities, energies, fluxes, and so on vary widely with both time and position (including altitude) in orbit. The lower-energy constituents of the space environment, notably neutral particles, plasmas, and fields, can be dramatically perturbed by the presence and operation of space systems, creating a local environment much different from the natural one.

The impact of system interactions must be examined in the context of the local space environment, leading to results that will, of course, be system-dependent. The higher-energy constituents of the space environment, such as radiation and particulates, are less influenced by the system than are the lower-energy constituents. Accordingly, the high-energy constituents can most readily be considered in terms of their direct impact on the system. However, to the extent that these constituents modify the system, their impact will also be manifested in the local space environment of lower-energy components.

Orbital Environmental Impacts

Two key factors in the operation of SDI systems in natural orbital environments drive system design in terms of both feasibility and launch weight. These are (1) tolerance of effluents, most dramatically those from open-cycle cooling and/or power systems; and (2) achieving satisfactory operation of very-high-power systems in the natural orbital environment. Effluents and high-power operation are clearly interrelated, and in the final analysis these factors must be considered at an overall system level, because the local space environment couples the impacts of various subsystems (e.g., the power and weapon subsystems). Furthermore, both the power levels and the effluent-expulsion levels envisioned for SDI systems are orders of magnitude beyond present experience in space. Resolution of these factors will require more than simple extrapolation of existing knowledge.

In evaluating possible impacts of effluent from open-cycle space power systems and space weapon systems, consideration must be given to effluent behavior under a variety of circumstances. Data from past experience in space will be of limited utility, however, since attitude-control jets produce effluent on a much smaller scale than will open-cycle space power systems. Those data do suggest that condensation may occur on cold surfaces of the spacecraft, and that there is also the possibility of creating "snow flakes," such as those reported in the Apollo program.

Chemically reacting gases or ionized gases could also emit electromagnetic radiation. For an SDI platform, such a radiant plume could interfere with its sensors, increase its detectability, and increase its vulnerability.

High-power systems use high voltages and/or high currents, hence operation of such systems in the low-pressure, ionized space environment requires great care to avoid surface arcing and "vacuum" breakdowns. The usual approach to operating high-power systems in the earth's atmosphere is to insulate the high-voltage components using oils or pressurized gas containers. Implementation of this approach for space power systems may make them prohibitively massive, especially since the mass problem is further aggravated by the need to make space systems survivable against natural and hostile threats: damage from space debris and meteoroids must be prevented over long mission lifetimes.

Survivability concerns introduce an additional factor that is best considered in synergism with dealing with effluents and providing high-voltage insulation. For example, if space platforms were hardened to several calories per square centimeter, the substantial armor shell employed would provide an environment that may eliminate the possibility of electrical breakdown between bare electrodes and at the same time protect against effluents and space debris. Thus, responding to survivability needs could also serve these two additional purposes. The prospect for simultaneously achieving these multiple objectives suggests that there should be an effort both to develop low-density, high-dielectric-strength materials for encapsulation and to formulate appropriate space experiments for testing them.

Arcing on partially insulated probes (planar solar-array segments) biased negatively in plasmas, both in ground test facilities and in orbit, has been observed at voltages in the range of a few hundred volts. Probes of other geometries—such as conducting discs on insulator and "pinhole" geometries—are less prone to arcing. While arcing mechanisms are not fully understood, enough is known that attempts to develop arc-resistant designs could prove fruitful. An initial demonstration of this possibility is the success of the SPEAR-I rocket experiment, which obtained data at altitudes up to 369 km. On December 13, 1987, using specially designed 1-m long booms to suppress surface arcing, two 20-cm diameter spherical probes on SPEAR-I demonstrated space vacuum insulation that withstood a maximum (pulse) voltage of 44 kV applied between the rocket body and the spherical boom terminal. The payoff, in terms of reduced power system mass, could be large if insulation requirements can be reduced or partially eliminated in pulsed systems.

The basic strategy is to achieve a design solution that both eliminates along-surface arcing and takes advantage of the space vacuum to help avoid gas breakdown when using high-voltage pulses. Steps along these lines are being taken within the current SDI program. However, it should be noted that attempts to follow this strategy may be compromised by high levels of contaminants, both because (a) surface arcing and gas-breakdown thresholds are dependent on background neutral and plasma densities, and (b) breakdown near a high-voltage surface can short out such space-vacuum-insulated conductors. Neutral-gas-breakdown threshold voltages are reduced below the Paschen levels when plasma is present (due to the availability of free charges for initiation), and—under some conditions—are further reduced by magnetic fields.

Magnetic fields at levels that may be present in high-power equipment may inhibit both charge mobility and breakdown potentials. In the density regimes characteristic of the unperturbed natural space environment, pressures are so low that very high voltages are necessary to initiate gas breakdown. Yet an orbiting vehicle introduces local surrounding gas and plasma densities through outgasing—as well as local magnetic fields characteristic of the orbital vehicle—creating conditions markedly different from those of the natural environment. Again, large amounts of effluent will make the system much more susceptible to breakdown and dictate more stringent insulation requirements. In addition, attaining survivability against hostile threats may complicate implementation of the exposed high-voltage approach, since direct use of the space vacuum as an external surface insulator will be precluded for those space platforms that must be hardened by encapsulation to withstand hostile threats.

In the SPAS studies examined by the committee, open-cycle space power systems are tentatively regarded as attractive choices compared to closed-cycle power systems because of their potential to be significantly less massive. As this report was being finalized, the committee became aware of a study by El-Genk et al. (1987) favoring closed-cycle nuclear power systems.

Open-cycle space power systems may not be practical if they liberate large amounts of effluent. Open-cycle weapon systems (e.g., chemical lasers) are also being proposed, although any resulting evolution of effluent clouds is not well understood in terms of expansion, dissipation, ionization, excitation, radiation emission, and interactions with surfaces, background environments, sensors, and weapons.

To understand the impacts of effluent clouds, estimates are required of neutral particle densities around a vehicle due to backstreaming around nozzles. Such estimates vary widely, however, depending on the nozzle geometry and the analytical models employed. There are orders-of-magnitude differences between independent predictions of densities in the region "behind" nozzles. None of the models has been fully validated. Impacts of effluents include chemical interactions with surfaces, condensation on cold surfaces, interference with power system operation, and interference with weapons and sensors. All of these impacts depend on the kind of effluent and its mode of emission, density, and temperature. The question of what liberated effluents, if any, can be tolerated has not been resolved. Yet the mass penalties resulting from containing large quantities of effluents are significant.

Various possible impacts of effluents on weapons and sensors need resolution. For example, one unresolved issue is whether effluent releases will interfere with propagation of a neutral particle beam. The simplest approach to estimating such interference is to approximate the effluent as a spherical cloud emanating from a point source at distance R_o from the particle beam source, then compute an effluent dump rate that will interfere with the beam. The validity of using the spherical approximation has been estimated by comparison to Space Shuttle data via the calculations shown in Appendix D. These calculations suggest both that the approximation is reasonable and that the issue of neutral-particle-beam stripping must be addressed. To complement these estimates, a program of space experiments is needed, as suggested in Recommendation 2.

From the space power system perspective, two strategies appear to have high payoffs in terms of reducing system mass. One is to explore development of effluent-tolerant systems; the other is to explore using space as a "vacuum insulator," if this approach is consistent with achieving survivability against hostile threats. The two are antithetical, as an effluent-tolerant system must be carefully insulated for the long term, while a space-vacuum-insulation approach may be intolerant of effluents. Given the prohibitive mass estimates generated in systems studies to date, it is imperative to quantify both of these potentially high-payoff mass-reduction approaches.

Because effluents can affect power systems, sensors, and weapons, analysis of the total space system must be significantly refined before a better understanding of the effluent issue can be reached. Meanwhile, aggressive pursuit of both the above strategies and space experimentation (Recommendation 2) appear to be indicated.

Resolution of both the high-power and effluent issues will require modeling and validation of models through comparison both with ground test data and with data from spaceflight testing. Meanwhile, the Space Plasma Experiments Aboard Rockets (SPEAR) program is obtaining flight data to address some of the high-voltage pulse-insulation issues. However, the SPEAR results may be experiment-specific, hence their generalized application requires caution.

Relationships among efflux density, background density, and velocity are such that a high-altitude, rocketborne experiment may be a good vehicle to obtain early data on the evolution of effluent clouds. These data would be applicable to high-altitude (greater than about 1,000 km) orbital vehicles. Because these issues are inherently total-system dependent, close cooperation among SDI's power, weapon, and sensor programs is indicated. Experience with the Space Shuttle (see Appendix D) can cast some light on impacts of effluent dump rates projected for SDI systems.

Conclusion and Recommendation

Based on the preceding, the committee arrived at the following conclusion and recommendation:

Conclusion 3: The amount of effluent tolerable is a critical discriminator in the ultimate selection of an SDI space power system. Pending resolution of effluent tolerability, open-cycle power systems appear to be the most mass-effective solution to burst-mode electrical power needs in the multimegawatt regime. If an open-cycle system cannot be developed, or if its interactions with the spacecraft, weapons, and sensors prove unacceptable, the entire SDI concept will be severely penalized from the standpoints of cost and launch weight (absent one of the avenues stated in Conclusion 2, Chapter 4).

Recommendation 2: To remove a major obstacle to achieving SDI burst-mode objectives, estimate as soon as practical the tolerable on-orbit concentrations of effluents. These estimates should be based—to the maximum extent possible—on the results of space experiments, and should take into account the impact of effluents on high-voltage insulation, space-platform sensors and weapons, the orbital environment, and power generation and distribution.

4
Needed Technological Advances in Space Power Subsystems to Meet SDI Requirements

IMPLICATIONS OF SDI SPACE POWER ARCHITECTURE SYSTEM STUDIES FOR ADVANCES NEEDED IN POWER SUBSYSTEMS

The following seven listings indicate key features of some space power systems selected by SDI space power architecture studies as being capable of providing power—in the relatively near term—at the levels indicated.

- At maximum power levels greater than 100 Mwe:
  1. open-cycle, gas-cooled fission reactor + turbine;
  2. open-cycle, $H_2 + O_2$ combustion + turbine;
  3. closed-cycle, Brayton, gas-cooled fission reactor + turbine; and
  4. closed-cycle, Rankine, liquid-metal cooled fission reactor + turbine.
- At power levels of 10MWe or less:
  5. closed Rankine cycle, fission reactor;
  6. closed Brayton cycle, fission reactor; and
  7. thermionic conversion, fission reactor.

These selections were provided to the committee in the form of prepublication results obtained from three simultaneous, independent studies of Space Power Architecture System (SPAS) options

(1988). These SPAS studies were performed under contract to the SDIO Power Program Office. It should be noted that the studies apparently did not make allowances for system survivability, leading the committee to Finding 3 below. Abbreviated descriptive summaries of some results of the SPAS studies are shown in Tables 4-1 and 4-2, and in Figure 4-1. These exhibits make no allowance for the mass of the hydrogen used for weapons cooling or for H_2-O_2 combustion.

The entries in Table 4-1 for masses of the open-cycle systems (first two columns) do not include the masses of hydrogen needed for cooling the reactor or for burning in the turbine. Rather, the table assumes the specific direction given to the SPAS contractors: namely, that the hydrogen for these purposes is available "free"—for example, with no mass penalty—from the weapon system. On the other hand, coolant required for the 1,800-s burst is included in the masses for the closed-cycle systems. According to the Sanda people who prepared Table 4-1 (Cropp, 1988), the turbine design for the H_2-O_2 combustion system was optimized assuming free hydrogen, so that simply adding the required mass of hydrogen will somewhat overestimate the overall system mass because of mass tradeoffs between hydrogen mass and turbine mass.

Another view of system mass comparisons is shown in Figure 4-2, from the IEG Field Support Team's critique of SPAS contractor reports. This figure shows an overview of the power system masses (exclusive of power conditioning) calculated by the SPAS contractors in terms of specific mass as a function of run time. The bands labelled "Open power systems," "Closed power systems," and "Closed thermodynamic cycle power systems" indicate the envelopes of contractor-calculated masses for these classes of systems. The mass of hydrogen is included in all of these specific mass figures, as is seen from the increases in mass with run time for all but the thermodynamic cycle systems. In Figure 4-2, the power systems that produce effluents that are discharged into space (which are referred to as open power systems) include (where TRW means TRW, Inc.; MM means Martin-Marietta, Inc.; and GE refers to the General Electric Co.):

- TRW Nerva-derivative reactor MHD (least massive)
- TRW gel MHD
- TRW Nerva-derivative reactor turboalternator
- TRW H_2-O_2 combustor turboalternator
- GE Nerva-derivative reactor turboalternator
- GE pebble bed reactor MHD

TABLE 4-1 Comparison of Masses of Burst-Mode Space Power Systems (500 MW, 1,800-s operation; masses are in metric tons)

Component	Power System						
	Open Gas-Cooled Reactor[a]	Open H_2-O_2 Combustion	500 W-h/kg Energy Storage	100 W-h/kg Energy Storage	1500°K Brayton Storage	1350°K Rankine	Thermionic with Thermal Storage
Power source	3.2	76.4	504.0	2,520.0	29.7	15.0	743.0
Shield	0.0	--	--	--	0.0	0.0	0.0
Turbine and generator[b]	31.8	43.3	--	--	27.0[c]	155.0[c]	--
Compressor	--	--	--	--	--	--	--
Radiator	--	--	24.0	24.0	1,468.1	549.5	--
Vapor separator	[illegible]					[illegible]	

Power conditioning	100.0	100.0	100.0	100.0	100.0	[illegible]	[illegible]
Power cond. and generator radiator	--	--	--	--	71.4	--	--
Miscellaneous	13.5	22.0	62.8	264.4	169.6	92.3	84.3
TOTAL	148.5	241.7	690.8	2,908.4	1,865.7	1,016.1	927.3

[a] Hydrogen is assumed available from the weapon-cooling system, hence its mass is not included in these masses.
[b] The specific mass of a generator is assumed to be 0.05 kg/kW for a standard generator. The mass of a cryogenic generator may be lower by a factor of two.
[c] Although some turbine mass differences are expected between Brayton and Rankine systems, these differences may be too large. Consistent algorithms for power-conversion system masses are still being formulated.
[d] The specific mass of power conditioning used here is 0.2 kg/kW. This is an average mass estimate that depends on the type of weapon system to be powered. Power conditioning for beam weapons may be somewhat more massive, but will be somewhat less massive for electromagnetic launcher weapon systems.

SOURCE: Sandia National Laboratories and NASA, Independent Evaluation Group Field Support Team, using reference models they developed prior to the Space Power Architecture System (1988) studies.

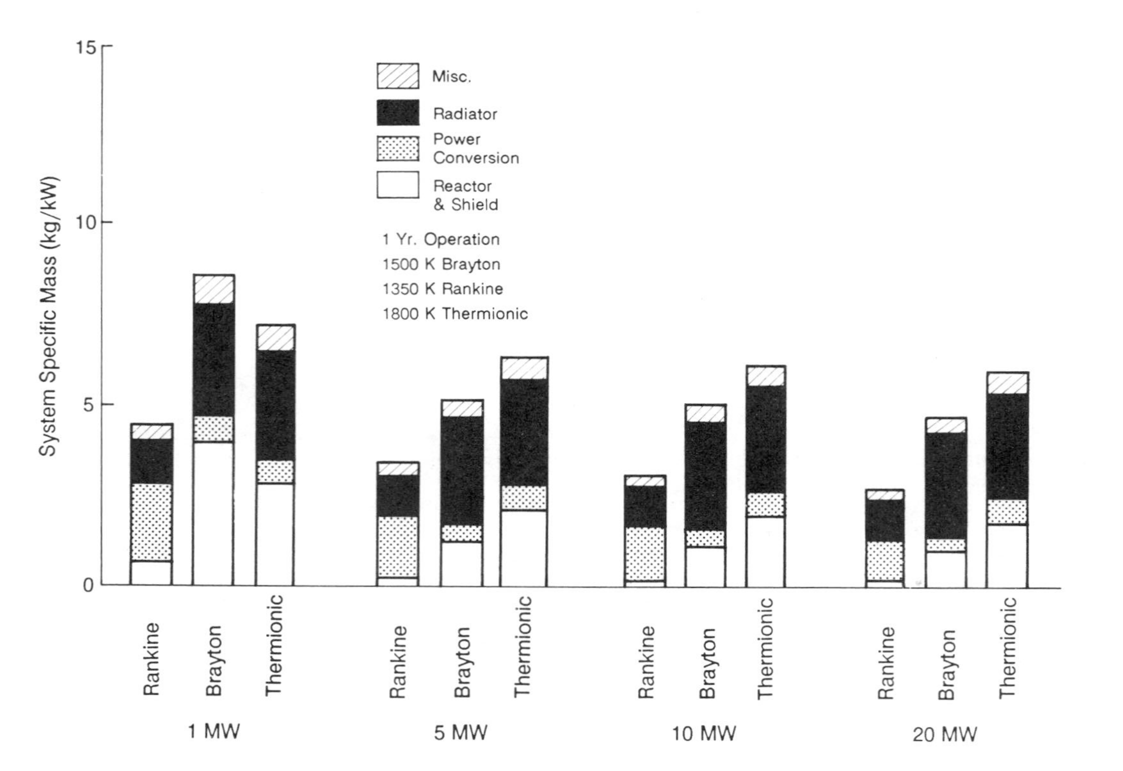

FIGURE 4-1 Comparison of specific masses of multimegawatt steady state-mode space power systems. NOTE: Reactor and shield information is inadequate to account for the variance between Rankine and Brayton systems. SOURCE: Sandia National Laboratories and NASA, Independent Evaluation Group Field Support Team, using reference models they developed prior to the Space Power Architecture System studies (1988).

TABLE 4-2 Multimegawatt Space Power System Comparison (10-MW, 1-year operation; masses are in metric tons)

Component	Power System		
	Rankine	Brayton	Thermionic
Reactor	1.1	5.5	12.8
Shield	1.1	4.6	6.6
Turbine and generator[a]	7.4[b,c]	0.8[b,c]	--
Compressor	--	0.5	--
Vapor separator	3.8	--	--
Power conditioning and generator radiator	1.4	1.4	2.0
Power conditioning	2.0	2.0	5.0
Radiator	11.4[b]	31.0[b]	30.0
Miscellaneous	2.8	4.6	5.6
TOTAL	31.0	50.4	62.0

[a] It is assumed that the specific mass of a standard generator is 0.05 kg/kW. The mass of a cryogenic generator may be lower by a factor of two.
[b] No allowance is made for weapons cooling.
[c] Although some turbine mass differences are expected between Brayton and Rankine systems, these differences may be too large. Consistent algorithms for power conversion masses are still being formulated.

SOURCE: Sandia National Laboratories and NASA, Independent Evaluation Group Field Support Team, using reference models they developed prior to the Space Power Architecture System (1988) studies.

- GE open H_2-O_2 fuel cell
- GE H_2-O_2 combustor turboalternator
- GE Li-HCl battery
- MM H_2-O_2 combustor MHD (most massive)
- MM H_2-O_2 combustor turboalternator
- MM open H_2-O_2 fuel cell
- MM combustor turboalternator (no H_2O)
- MM Nerva-derivative reactor turboalternator

The power systems that generate chemical products that are retained (known as closed power systems) include

- TRW ice-cooled H_2-O_2 fuel cell (least massive);
- MM ice-cooled H_2-O_2 fuel cell with radiator;
- TRW closed combustor turboalternator (most massive); and
- TRW lithium-metal sulfide battery.

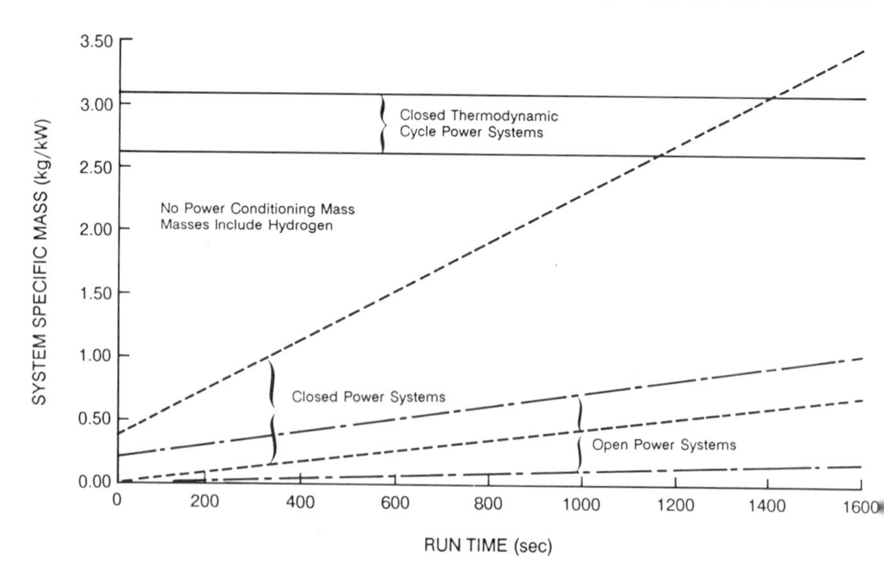

FIGURE 4-2 Mass as a function of whether the system is open or closed and of run time. SOURCE: Sandia National Laboratories, Independent Evaluation Group Field Support Team, based on inputs from Space Power Architecture System contractors (1988).

The closed thermodynamic cycle systems are Martin Marietta's reactor-powered Rankine and thermionic systems, which use radiators to reject waste heat.

Figure 4-3 is a block diagram showing a closed-cycle Brayton space power system energized by a nuclear reactor. Figure 4-4 is a similar diagram for an open-cycle space power system energized by the combustion of hydrogen and oxygen. Note that weapons cooling is diagrammed in Figure 4-4.

The above-mentioned descriptive summaries are based on three SPAS studies performed by SDIO-supported contractors who, unfortunately, employed an inconsistent set of assumptions. Consequently, there are necessarily differences in the three sets of results that are difficult to interpret, a problem recognized by the SDIO Power Program Office's technical team charged with interpreting the SPAS results. This problem is especially severe in comparing estimates of system mass. In that regard, this team noted significant differences in assumptions among the contractors—along with overall limitations in the assumptions—pertaining to the following technological and packaging considerations, to which the mass estimates are sensitive:

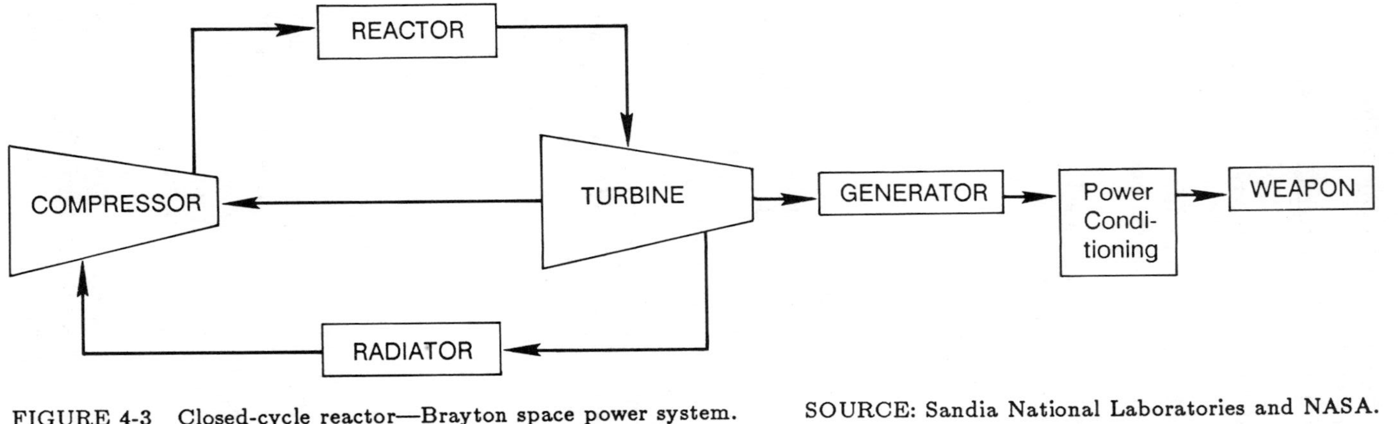

FIGURE 4-3 Closed-cycle reactor—Brayton space power system. SOURCE: Sandia National Laboratories and NASA.

• High-voltage power systems perform well with tube radiofrequency (RF) systems.

• Low-voltage power systems perform well with solid-state RF systems.

• High-voltage alternators save mass, since no transformers are needed.

• Cryo-cooled power conditioning, if realizable, saves mass, hence conductors, transformers, and other components can be less massive.

• Mass estimates were based on conservative near-term or on optimistic far-term assumptions regarding technology.

• Masses required for thermal management and packaging were not uniformly considered.

• The technology postulated for power conditioning does not exist.

Tables 4-1 and 4-2 warrant comment, in view of the fact that the information they contain became the bases for several of the committee's findings, conclusions, and recommendations. In Table 4-1, masses for several burst-mode space power systems are quoted in metric tons. For the convenience of those unaccustomed to thinking in those units, Table 4-3 shows the range of system masses from smallest to largest.

Assuming typical costs per pound for development, production, and launching to orbit, and noting that the power system may range from 20 to 50 percent of the total orbital vehicle mass, these systems appear to be very large—hence probably prohibitively expensive—and too massive to lift into orbit with any practical launch vehicle, unless they were launched separately and assembled in orbit, thus motivating Conclusion 2 below.

A further difficulty encountered in Table 4-2 is the significant difference in reactor masses between the Brayton and Rankine systems. This large discrepancy resulted from the fact that the two sets of results were obtained by separate contractors who used different technical assumptions, some of which may be questionable. The range of their results for reactor and turbogenerator masses of two multimegawatt space power systems is shown in Table 4-4.

These differences contributed significantly to the committee's reluctance to recommend, with any assurance, either the selection or elimination of any candidate space power system(s). Figure 4-1 expresses the above results in terms of system-specific masses (in kg/kWe), for gross power levels of 1, 5, 10, and 20 MWe. In this

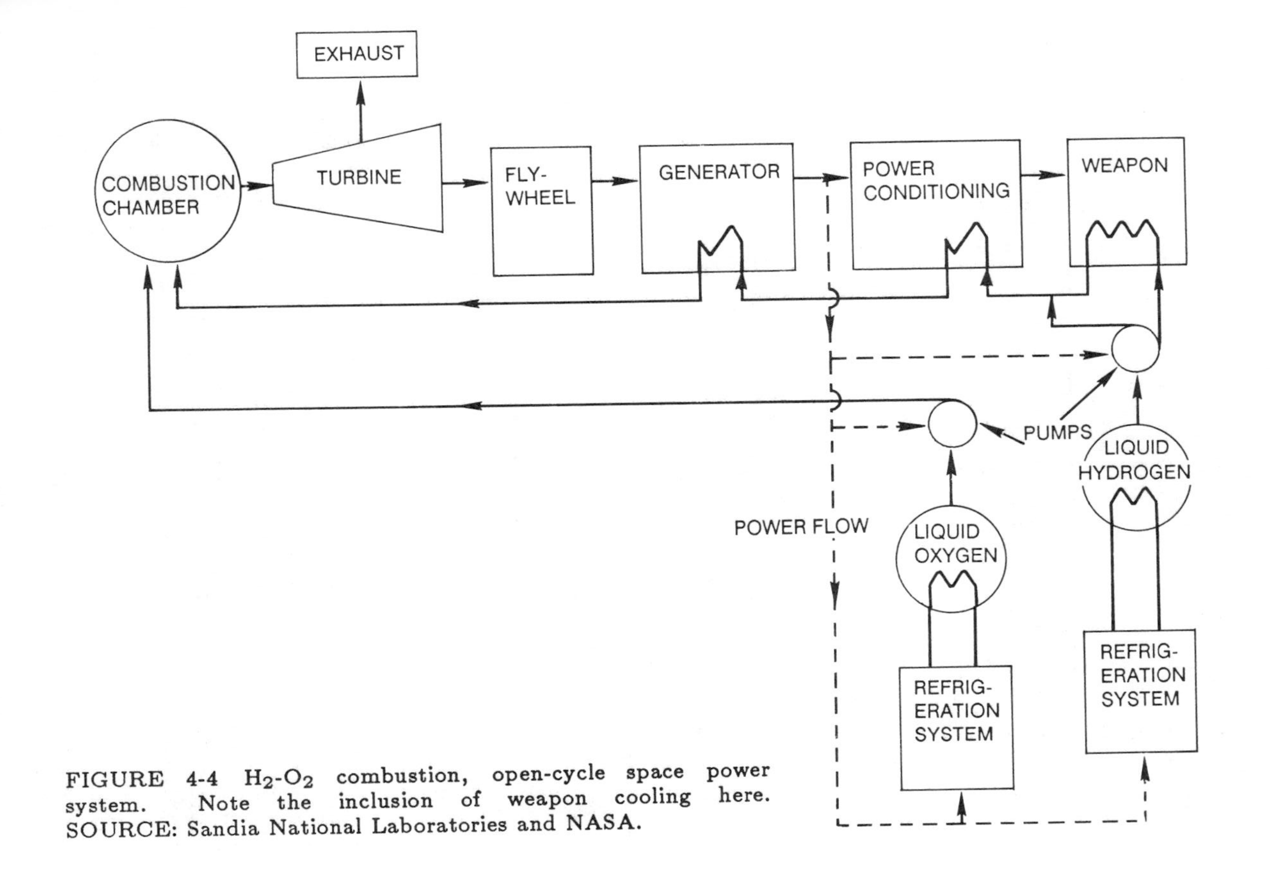

FIGURE 4-4 H_2-O_2 combustion, open-cycle space power system. Note the inclusion of weapon cooling here. SOURCE: Sandia National Laboratories and NASA.

TABLE 4-3 Range of System Masses of Various SPAS Burst-Mode Space Power Systems

System	System Mass		
	Metric tons	Kilograms	Pounds
Smallest	148.5	148,500	326,700
Largest	2,908.4	2,908,400	6,398,480

NOTE: For comparison, the payload capacity of the Space Shuttle is currently about 45,000 pounds, and the largest U.S. heavy-lift launch vehicle could lift less than the 326,700 pounds stated above. SPAS = Space Power Architecture System.

TABLE 4-4 Range of Reactor and Turbogenerator Masses of Two SPAS Multimegawatt Space Power Systems

System	Reactor Mass		
	Metric Tons	Kilograms	Pounds
Reactor (Rankine)	1.2	1,210	2,662
Reactor (Brayton)	5.5	5,500	12,100
Turbine and generator (Rankine)	7.4	7,400	16,280
Turbine and generator (Brayton)	0.8	800	1,760

NOTE: SPAS = Space Power Architecture System.

form, the difference between Rankine and Brayton cycles is open to question. Perhaps most striking is the difference between these cycles for reactor plus shield, a factor of 5 to 6 at all gross power levels.

Figure 4-3 indicates, for one particular system, the integration of power system and weapon—at least to the extent of using a common source of hydrogen. In other systems studied, there is no explanation of the extent to which the power system was reviewed in the context of the entire orbital vehicle. Consequently, the committee cannot assess either penalties or advantages that might be encountered in making the power system an integrated part of the complete orbital platform.

These difficulties (Findings 2, 3, and 4) lead the committee to the conclusion that there is still an insufficient basis for making a selection between the power architectures examined; they also motivate the committee's Recommendation 1—to carry out detailed, whole-system design studies.

Despite difficulties in comparing the SPAS results, however, the dramatic mass differences between open-cycle and closed-cycle power systems, found by all of the contractors (cf. Fig. 4-2), were qualitatively adequate to motivate Recommendation 2 regarding the question of permissible effluent. Figures 4-3 and 4-4 highlight the differences between typical closed-cycle and open-cycle systems, respectively.

ADVANCES NEEDED IN HIGH-TEMPERATURE STRUCTURAL MATERIALS TECHNOLOGY

As is evident in Tables 4-1 and 4-2, radiator masses are a large fraction of the mass of closed-cycle (Brayton and Rankine) space power systems. As the peak operating temperature (T, measured in °K) of a space power plant increases, the heat radiated per unit radiator mass increases rapidly (as T^4). Therefore, advances in current materials technology could provide high-temperature, creep-resistant materials that could greatly reduce the radiator mass required (Rosenblum et al., 1966; Buckman and Begley, 1969; Devan and Long, 1975; Klopp et al., 1980; DeVan et al., 1984; Stephens et al., 1988).

For example, if the technology for carbon–carbon composites were sufficiently advanced so as to provide a material for constructing a Brayton cycle power plant, it is conceivable that the turbine-inlet temperature could be raised from the 1500°K stated in Table 4-1 to 2000°K. This temperature increase would reduce the mass of the required radiator by about a factor of three, thereby roughly halving the total mass of such a power system, and would also increase the efficiency of power conversion. Realization of the full potential of such a material, about 2300°K (National Research Council, 1988), would reduce power system mass even further. Accordingly, achieving broadly based advances in high-temperature structural materials could provide a basis for dramatic potential gains in power plant performance and corresponding reductions in power plant mass. Use of such materials in space would avoid the need for the antioxidation coatings that are required for terrestrial applications in an oxidizing atmosphere.

TABLE 4-5 Potential Department of Defense (DOD) Applications of Superconductors for Power Components

Power Application	Responsible DOD Organization(s)
Megawatt power generation (low specific mass: 0.1 kg/kWe)	
Synchronous alternators	Air Force, SDI
Pulsed alternators	Air Force, Army, BTI, DARPA
DC generator exciters	Air Force, Army, BTI, DARPA, Navy
MHD generator magnets	Air Force, BTI
Megawatt propulsion motor (DC)	Navy
Power conditioning and energy storage	
Low-mass, fast-pulsed energy storage	Air Force, Army, BTI, DARPA, SDI,
Ground-based, slow-pulsed energy storage	Air Force, Army, DNA, SDI
Low-mass inverter transformer	Air Force, SDI
Low-mass inductor components	Air Force, DNA, SDI
Power transmission lines	Air Force, Army, BTI, DARPA, DNA, Navy, SDI

NOTE: BTI = Balanced Technology Initiative; DARPA = Defense Advanced Research Projects Agency; DC = direct current; DNA = Defense Nuclear Agency; MHD = magnetohydrodynamics; SDI = Strategic Defense Initiative

ADVANCES NEEDED IN POWER-CONDITIONING AND PULSE-GENERATING TECHNOLOGIES

Superconducting Materials

Superconductors are potentially useful throughout the power system/weapon system. The importance of superconductors in power applications lies in their ability to carry large current densities with essentially no resistive losses. Tables 4-5 and 4-6 list potential SDI power- and weapons-related applications, respectively.

In view of their potential to operate in liquid hydrogen, the recently discovered high-critical-temperature superconductors could impact many SDI applications if they can be developed into usable forms. As a result of this potentially major impact, a more detailed discussion on these materials is provided in Chapter 5. Increased research in this area is being sponsored by industry, the Department of Defense, Department of Energy, and the National Science

Foundation. These agencies are redirecting funding into high-critical-temperature superconducting materials and their applications.

Component Technology

The state of the art in power conditioning is adequate to satisfy the needs of most commercial land-based power applications. For those applications, it is sufficient to improve product reliability without developing new devices. For instance, radar designs, both airborne and land-based, are rather standard and use only proven components, resulting in the availability of only a few competing designs. This is done to avoid the cost of developing new components and the subsequent need for a program to prove their reliability. The same

TABLE 4-6 Potential Department of Defense (DOD) Applications of Superconductors for Weapons Components

Weapon Application	Responsible DOD Organizations
Directed Energy	
Laser	
RF cavities	Air Force, Army, SDI
Wiggler magnets	Air Force, Army, SDI
Electron beam guidance magnets	Air Force, Army, SDI
Particle beam	
RF cavities	Air Force, Army, SDI
Beam-guiding magnets	Air Force, Army, SDI
Focusing magnets	Air Force, Army, SDI
Kinetic Energy (electromagnetic launchers)	
Tactical	
Augmentation magnets (railguns)	Army, BTI, DARPA, DNA
High-current switches	Air Force, Army, BTI, DARPA, DNA
Coil gun accelerators	Army, BTI, DARPA
Strategic	
Augmentation magnets (railguns)	Air Force, DNA, SDI
High-current switches	Air Force, SDI
Coil gun accelerators	Air Force, SDI

NOTE: BTI = Balanced Technology Initiative; DARPA = Defense Advanced Research Projects Agency; DNA = Defense Nuclear Agency; RF = radio frequency; SDI = Strategic Defense Initiative.

can also be said of much of the electronics associated with launch vehicles and satellites.

Newer applications requiring very fast pulses or very high average powers have met with difficulties, in that the present state of the art in component technology is generally inadequate to achieve the desired level of performance (Rohwein and Sarjeant, 1983). These applications have not offered sufficient economic impact to stimulate substantial corporate investment in a new technology base required to establish the next generation of power-conditioning designs. Instead, such applications have attained their rather modest goals through modifications and extensions of existing components or techniques.

Although present designs serve very well, they do not scale directly into the multimegawatt range for SDI applications. This difficulty is partly attributable to the emphasis placed on conservative designs in order to obtain the requisite reliability; however, in the multimegawatt range, extension of standard designs leads to impractically large and massive systems. More must be known about the failure mechanisms of critical components. The mass penalty of the large design margins affordable in small systems cannot be tolerated at high SDI power levels. Indeed, entirely new components and concepts may be required to achieve SDI objectives. Power must be made available at specific voltage and current levels matching weapons power requirements.

FINDINGS, CONCLUSION, AND RECOMMENDATION

Based on the discussions in this chapter, the committee arrived at the following findings, conclusion, and recommendation.

Finding 2: The space power subsystems required to power each SDI spacecraft are a significant part of a larger, complex system into which they must be integrated, hence the only completely valid approach is to analyze them in the system context. (see Conclusion 2 and Recommendation 1.)

Finding 3: Existing space power architecture system studies do not adequately address questions of survivability, reliability, maintainability, and operational readiness—for example, availability on very short notice.

Finding 4: Existing SDI space power architecture system studies do not provide an adequate basis for evaluating or comparing cost or cost-effectiveness among the space power systems examined.

Conclusion 2: Gross estimated masses of SDI space power systems analyzed in existing studies appear unacceptably large to operate major space-based weapons. At these projected masses, the feasibility of space power systems needed for high-power SDI concepts appears impractical from both cost and launch considerations. Avenues available to reduce power system costs and launch weights include (a) to substantially reduce SDI power requirements; (b) to significantly advance space power technology.

Recommendation 1: Using the latest available information, an in-depth full-vehicle-system preliminary design study—for two substantially different candidate power systems for a common weapon platform—should be performed now, in order to reveal secondary or tertiary requirements and limitations in the technology base which are not readily apparent in the current space power architecture system studies. Care should be exercised in establishing viable technical assumptions and performance requirements, including survivability, maintainability, availability, ramp-rate, voltage, current, torque, effluents, and so on. This study should carefully define the available technologies, their deficiencies, and high-leverage areas where investment will produce significant improvement. The requirement for both alert-mode and burst-mode power and energy must be better defined. Such an in-depth system study will improve the basis for power system selection, and could also be helpful in refining mission requirements.

5
Approaches Toward Achieving Advances in Critical Power Technologies

In the discussion of space-based power requirements (Chapter 2), the committee pointed out the advantage of pursuing high-leverage areas; similar approaches can yield some very useful results in advancing critical power technologies. In this chapter, the following subjects are discussed: advancing thermal-management techniques, advancing power-conditioning components and technologies, and materials advances required for developing power component technologies.

ADVANCING THERMAL-MANAGEMENT TECHNOLOGIES

The thermal-management problem is that all heat generated on a space platform must be (a) converted into another form of energy (with the associated thermodynamic constraints); (b) absorbed as temperature rise in components or thermal storage elements; (c) absorbed by a coolant that is vented; or (d) radiated to space either directly from the component or by use of a higher-temperature, more efficient radiator.

The last option requires a heat-pump (refrigeration) cycle, in which heat is absorbed at a low temperature and rejected by the radiator at a higher temperature. Only the first three options are available for heat rejection from the space power system itself.

The space power system—defined for this purpose to include

a heat source, power conversion devices, and its loads—is the primary source of spacecraft-generated thermal energy that must be disposed of. Thus the efficiencies and losses of the overall power system—including those of its subsystems and components—are major factors determining how much heat is generated and thus must subsequently be absorbed or rejected. Availability of survivable, cost-effective technology to store, pump to higher temperatures, and radiate thermal energy effectively with low mass penalties is an important ingredient of space power system design.

The problem of thermal management is very important for spacecraft of any size, to say nothing of spacecraft power systems ranging from hundreds of kilowatts to multimegawatts. The primary means for heat rejection currently employed is to use heat radiators. This method is basically the only long-term means of rejecting heat in space without spacecraft mass alteration. Obviously, heat can be stored in a mass that is then ejected from the spacecraft. The practicality of this method is limited (to about 30 min) by the rapid increase of the mass required with increasing duration of operation. Further, heat storage (in a heat sink) is a very useful method of point cooling and has considerable potential for SDI utilization. These methods will be discussed separately.

Heat-Rejection Considerations

As is well known, the amount of heat radiated from a surface is proportional to the fourth power of the surface temperature (measured in °K) and to the emissivity of the surface material. For these reasons, reductions in radiator size and mass can be realized if the operating temperatures and emissivities of space power radiators can be increased. Because of the high sensitivity to temperature, dramatic mass reductions can be achieved, as discussed in Chapter 4, whereas there is less sensitivity to emissivity improvements.

Significant innovation in this area has the potential to alter conventional views of power-system design trade-offs and should be examined in connection with the preliminary vehicle design proposed in Recommendation 1 of this report. Innovative radiating systems based on liquid-droplet radiators, moving-belt radiators, heat pipes, or on radiators that are deployed on power demand have been proposed. Although there is no assurance that any of these concepts will prove feasible, such approaches might produce significant reductions in radiator size and specific mass, and hence warrant exploratory research. An unarmored deployable radiator would be less massive than an armored radiator, yet still be survivable to attrition attacks by ground-based or space-based lasers.

For high-power systems, needs for heat rejection and mass minimization cause system designers to favor power systems that operate at high temperatures, thereby reducing the size of power-conversion equipment (through higher conversion efficiencies), the amount of heat that needs to be rejected, and the size of the radiators. For low-temperature power systems, low-density materials (e.g., aluminum, beryllium, or titanium) can be employed as radiators, thereby providing a means for reducing mass. Unfortunately, most highly developed heat-rejection technology was optimized based on cost factors, rather than on considerations such as survivability, efficiency, or high-temperature capabilities. Consequently there is only a limited available technology base applicable to the problem at hand.

Heat rejection is essential, and for closed-cycle (noneffluent) space power systems at the multimegawatt level, the heat rejection subsystem (see Chapter 4, Figure 4-1) can easily account for half the mass of the overall power system. The SP-100 power system has large, massive radiators because of the low conversion efficiency of its thermoelectric converters. These radiators are made even more massive by the imposed survivability requirements. Two other heat rejection options are discussed below which avoid using radiators but are mass-intensive, hence they become impractical as the duration of power usage increases beyond about 1,000 s.

One of these options is to use heat storage aboard the spacecraft for thermal management of multimegawatt systems that are operated for only short periods of time (i.e., in the burst mode). Most of the heat storage needed can be accomplished through endothermic chemical reactions, use of specific heat capacity, and phase changes.

The other option that avoids radiators is gross heat rejection from a thermal engine, where the waste heat is simply thrown overboard with the effluent. This is a viable concept if the effluent does not unduly interfere with friendly weapon, sensor, spacecraft, or power systems. On the other hand, liberating effluents may hinder hostile action.

From a strategic standpoint, duration constraints on the use of the above mass-intensive options make them ineffective against a counter-strategy of prolonging the period of combat to an hour or longer.

Power system working fluid can typically be used for weapon

cooling prior to entering the power-generation system. The ejected effluent will thus have served the dual function of disposing of waste heat from both weapon and power-generation systems. Resolution of the question of whether or not the release of effluents is tolerable is addressed in Recommendation 2.

Survivability Considerations

The survivability of space radiators is a major design problem, owing to the ease of detection of such localized thermal sources. This problem is especially serious at the high rejection temperatures that might be used for nuclear reactor systems. Radiators at those temperatures could act as infrared homing beacons for hostile detection and action. Early in the process of advancing candidate space power systems, thermal rejection techniques need to be identified that mitigate the risk of detection and attack but do not impose excessive mass penalties for hardening. Candidate heat-rejection techniques showing promise should then be subjected to feasibility studies and scalability validation.

ADVANCING POWER-CONDITIONING COMPONENTS AND TECHNOLOGIES

Advances in power system components, materials, and technology are necessary to meet envisioned SDI requirements, as discussed below.

Advancing the Design of Conductors

Conductors usually make up a significant fraction of the overall mass of a power device and also determine its characteristics. Conductor mass is generally traded off against electrical losses, device efficiency, conductor temperature rise, complexity of cooling, and the amount of coolant required.

Normal Conductors

Practical conductors at ambient temperatures generally consist of copper or aluminum or their alloys. The materials are selected to give the appropriate combination of low resistivity, mechanical strength, and ease of fabrication required for specific applications.

High-strength conductors are important in applications (generally circular or solenoidal windings) where the conductor is also the only part (or a major part) of the necessary structure. In applications where the structure is separate or is not of major concern, the resistivity of the conductor becomes the major factor.

It is desirable to have conductors that can operate at high current density—consistent with achieving structural, dielectric, and thermal requirements of the winding.

Essentially all pure metallic elements of interest as conductors exhibit decreasing resistance with decreasing operating temperature. This decrease is limited at the low-temperature extreme by impurities, magneto-resistance, mechanical stress level, work hardening, size, and so on.

High-conductivity, high-strength, wide-temperature-range metallic conductors are distinctly possible, but do not appear to have been examined over the temperature ranges of interest. The two best metallic conductors (commercial, at practical cost levels, formable, ductile, and tough) are copper and aluminum and oxide-dispersion-stabilized (ODS) alloys of Cu and Al.

Dissolved impurities have major negative effects, especially on electrical conductivity, but also on thermal conductivity. The purer the Cu and Al, the higher the conductivity values. Fortunately it is possible (commercially) to produce Cu at purity levels of 99.9 percent or better, and Al at 99.99 percent or better. Each dissolved impurity element has a different effect (percent conductivity loss per unit of impurity content) on conductivity.

The potential availability of liquid hydrogen as a conductor coolant can have a major effect on system operating temperatures. Conductors capable of operating with liquid hydrogen should be developed either separately or as part of the component development.

Of existing conductors, high-purity aluminum in a composite with an aluminum alloy is a promising candidate for use in a liquid-hydrogen high-power alternator. If estimates of conductor performance hold up for this option, it could be important in many direct-current and alternating-current applications.

Superconductors

Superconductors exhibit zero resistance only below a certain critical temperature. They are poor electrical conductors above this value. The metallic superconductors NbTi and Nb_3Sn, which require liquid helium for their operation, are capable of achieving operational winding current densities of 50,000 A/cm^2, but are not suitable for operation at temperatures other than in the liquid helium range (about 4°K). Because of their potentially high current density, these alloys will continue to be major technological contenders for SDI applications.

For radiofrequency (RF) operation at low magnetic fields, superconductors exhibit low surface resistivities—the lower the temperature, the lower the losses. Nb cavities are being used at 2°K for accelerators in the gigahertz range. Since operation at radio frequencies is a surface phenomenon, a layer of superconducting material of the appropriate thickness must be carefully applied to a suitable substrate.

The use of superconductors in power systems generally leads to high-efficiency, compact components and subsystems. High efficiencies of power generation, power transmission, and power conditioning have direct beneficial effects on the ratings and masses of prime power sources and, in addition, a low-loss RF cavity reduces mass requirements for the entire system.

At present, most superconductor applications are either direct current (DC) or quasi-DC, where the currents change slowly. It is only during DC operation that superconductors have zero resistance to the flow of electrical current. The limits of this region of zero resistance are a function of operating temperature, current density, and magnetic field.

For alternating current (AC) operation at 60-Hz power frequencies, superconductors exhibit losses above a threshold magnetic field. These losses decrease with decreasing filament size—one of the reasons for the multifilamentary configuration currently being used in Nb-based superconductors. Until now, because of the penalty of having to operate at liquid-helium temperature, applications at power system frequencies have been limited by the unavailability of conductors having micron-size filaments. Achieving higher operating temperatures for these materials means reduced refrigeration requirements, and the prospect of doing so suggests that it is timely to review potential space power applications (at frequencies much greater than 60 Hz), such as transmission, transformers, inductors, armature windings, and so on.

Superconductor transition temperatures up to 125°K have now been reported by several institutions. There is no reason to believe that the optimum materials have been discovered, and further progress is expected. Reports of achieving critical temperatures of room temperature or above have been erratic, unconfirmed, or have used inadequate measurement techniques.

To summarize the present status of superconducting materials:

1. High-critical-temperature superconducting materials have the potential of carrying high critical currents.
2. The superconducting materials are ceramics and, at this stage of their development, they have poor mechanical properties.
3. Indications of transition temperatures above room temperature have been reported (Materials Research Society, 1987), but have not been confirmed and may not be reproducible.
4. Applications depend on the availability of superconductors or superconductor sections with consistent properties that can be fabricated into reliable windings for magnetic components or structures resembling permanent magnets.

This committee concludes that high-critical-temperature superconductors may well play a major role in SDI power applications someday. Nevertheless, because of their early stage of development, such superconductors are not currently available—nor will they likely be available for many years to come—to replace present or improved power technology. Accordingly, the development of other SDI power technology should not be curtailed until these superconductors begin to become a viable option.

Superconducting Magnetic Energy Storage

Storage of energy in a magnetic field occurs when electricity flows through one or more coils. Since any electrical resistance in the circuit causes energy loss, the use of superconducting coils—which have no DC resistance—is a very efficient approach to storing electrical energy for any length of time.

A major application of energy storage is to allow energy sources to be sized for average or low power. In the case of superconducting magnetic energy storage (SMES), the coils are energized at low power levels and then discharged at a higher power level.

Low-critical-temperature superconductor technology has been demonstrated on several large-scale projects—primarily in magnetic fusion, high-energy physics, and magnetic resonance imaging applications. Technology for these applications usually operates reliably, and even larger-scale applications of superconductivity in these areas are planned.

Studies aimed at providing ground-based power of limited duration at gigawatt power levels with rapid rise-times have indicated that SMES is a very attractive approach for SDI applications, especially in view of the possibility of time-sharing the facility with a utility during peacetime. Two large contracts for independent conceptual design studies of SMES systems were awarded by SDIO late in 1987.

ADVANCEMENT POTENTIAL OF TECHNOLOGY FOR DYNAMIC POWER-CONVERSION CYCLES*

The existing gas-turbine industry builds gas turbines for propelling aircraft and builds both gas and steam turbines for terrestrial power generation. The largest gas turbines generate 100 to 200 MWe per module, and specialized gas turbines operating on stored compressed air produce up to 290 MWe from a single machine (Gas Turbine World, 1987). About 20 separate models of gas turbine power plants currently marketed have power ratings exceeding 100 MWe.

Advances in the gas turbine for solar-dynamic power generation aboard the Space Station and for use with nuclear reactors such as SP-100 could occur (English, 1987) in the following ways:

- Using a tantalum-based refractory metal alloy (ASTAR-811C) for the hot components of the power plant would permit operation at peak temperatures up to 1500°K. That alloy has been creep-tested for over 300,000 hours at temperatures from 1144°K to 1972°K (Klopp et al., 1980). Refractory alloys—based on molybdenum and niobium—having considerably lower density may in the future prove to be applicable at these temperatures; specifically, the Mo-HfC alloy has been tested for only a few hundred hours at temperatures up to 1800°K in an inert gas atmosphere, much less than the testing (over 22,000 hours) to which ASTAR-811C was subjected at temperatures above 1800°K (Klopp et al., 1980). However, both molybdenum and niobium alloys must still undergo a very considerable testing program before final conclusions can be drawn.

---

*The committee has discussed the Brayton cycle in considerable detail. Many of the advances described in this section are also applicable to the Rankine cycle. The committee believes additional study of both cycles is warranted in view of unexplained or inconsistent SPAS analysis results, which were unavailable in published form during the course of this study.

- By using the Brayton cycle combined with molten-lithium heat storage, since the sensible heat capacity of the molten lithium is higher—by a factor of two or more—than the latent heat capacity of the fusible salts now contemplated for the Space Station. Use of lithium, because of its extremely favorable heat-transfer properties, would also permit a significant reduction in the size and mass of the solar heat receiver.
- Inasmuch as molten lithium is not tied to any given working temperature (as is the melting and freezing of a salt), using lithium in a Brayton cycle would permit the gradual evolution of a given power plant by first operating it at, say, 1200°K and then gradually raising the operating temperature toward the potential of the power plant, 1500°K in this case.
- This rise in peak temperature would increase not only the power generated but also the efficiency of power generation; the sizes of the solar collector and waste-heat radiator could therefore remain constant with up to a 50 percent increase in generated power.
- Finally, for application to nuclear power, the solar mirror and solar heat receiver of the solar Brayton power plant could be replaced by a lithium-cooled nuclear reactor, such as the SP-100.

By virtue of their high efficiency, closed Brayton and Rankine cycles could generate about 500 kWe using the same reactor from which the present SP-100 thermoelectric conversion design generates 100 kWe. Similarly, from a 25-MW reactor required for thermoelectric generation of 1 MWe, these power cycles would generate up to about 5 MWe.

For generating very high power in the burst mode, use of molten lithium as the heat sink for a high-power closed-cycle system would provide a low-mass power plant that discharges no effluent during operating periods of 1,000–2,000 s. This same technology could also provide the megawatts of power needed for long periods in the alert mode.

Advancement Potential for Alternator Technology

Alternators are electrical rotating machines that convert shaft energy into AC electrical power that can then be used as generated or transformed and/or rectified as required by the load. A field winding—usually DC-energized—is rotated, with the power being generated in the stationary armature.

The power for a given-size machine generally increases with

increasing speed and current density in the field and armature windings—within the limits of structural integrity—consistent with the requirements for high rotational speeds and rapid start-up sometimes imposed. The alternator technology relies heavily on the available prime mover, the available conductors, and thermal management of the losses in the rotor and stator.

For an ambient-temperature application, the U.S. Army is developing a 3-MWe, gas-turbine-driven, oil-cooled machine. The alternator for that device has a specific mass of about 0.1 kg/kWe (for the generator alone) and rotates at 10,500 to 15,000 rpm. The output power, which has a frequency of about 1 kHz, is fed into transformers.

Because superconductors have losses when subjected to time-varying currents or magnetic fields, the use of superconducting technology has been limited up to now to the field windings of the alternator, where they are exposed essentially to DC operation. An example of this technology is a machine using a liquid-helium-cooled superconducting rotating field and an ambient-temperature armature, being developed by General Electric for the U.S. Air Force The machine is undergoing preliminary testing. It has a rating of 20 MWe at 6,000 rpm and is capable of starting up in 1s from a cooled-standby condition. The machine has a specific mass of 0.045 kg/kWe, and is designed with several system-oriented unique features, such as a rectified 40-kV DC output, potentially eliminating the need for additional transformers. It also has an ambient-temperature aluminum shield that reduces external time-varying magnetic fields, which is an important design feature for space applications. Because of its lower speeds and high-voltage winding, the 0.045 kg/kWe machine is not directly comparable to the 3-MWe army machine.

An experimental air-core alternator with a disc rotor is being designed by ARDEC-KAMAN with a continuous rating of 0.1 kg/kWe at about 5,500 rpm. This rating is projected to decrease to 0.03 kg/kWe for a future cryogenic machine with counterrotating discs and a 20-MWe rating.

An approach using liquid-hydrogen-cooled, high-purity aluminum conductors for both field and armature is being undertaken for SDI by Westinghouse and Alcoa. Recent measurements of the resistivity of high-purity aluminum samples by these organizations (Billman, 1987; Eckels, 1987) are lower than previously attained. If such resistivity can be maintained in finished windings, these results indicate that high-purity aluminum may be an even better conductor than previously thought for high-current-density operation in the liquid hydrogen range. Estimated specific masses are of the order of 0.03 kg/kWe for a 30-MWe machine with output in the 50 to 100 kV range.

In the United States there is essentially no operating experience with alternators other than at ambient temperatures. Cryogenic and superconducting techniques have been successfully demonstrated in homopolar types of machines and in other stationary applications, such as magnets for high-energy physics, magnetic fusion experiments, and magnetic resonance imaging. While experimental developmental hardware does exist, the successful application of these techniques to high-power alternators still remains to be demonstrated.

An alternator configuration for use in space—because of its interface with power conditioning/load, thermal management, the prime mover/energy source, and the torques, magnetic fields, high voltages, and currents it generates—must be the result of a thorough, interactive systems approach. The basic advantages of the alternator in being able to generate high voltages without transformers must be traded off against the loss of flexibility in initially developing a general purpose alternator that must then be connected to a power system with transformers to provide load-specific voltage levels. Note that all loads may not operate at the same voltage level.

Direct generation of high voltages requires either placement of the alternator near the load or the transmission of power at high voltages, and has the attendant problems of high voltages in space and fault management, as discussed elsewhere in this report.

Advancing the State of the Art in Power System Components

Funding that has been available for component development has generally been used within a program to improve existing manufacturing techniques, evaluate new materials developed for other applications, and to make improvements in techniques for manufacturing components. In view of the limited resources available in the past, that was the only logical approach. However, this strategy is at best capable of achieving only modest gains.

A more cost-effective approach is illustrated by the example of the recent joint SDI/DNA capacitor program. This program has been very successful, in large measure because it made maximum use of new theory and computer modeling power. This program makes use of 1980s technology rather than simply extending the standard approach typical of the 1960s. This approach can be applied to other component technologies as well. Examples of the rapid advances achieved to date in representative power technology components when their development was aggressively funded are illustrated in Table 5-1. In contrast, note the dismal evolutionary advance rates of 1.5 per decade in surface-voltage withstand-level for resistors. While the committee recognizes that technical progress is often nonlinear, use is made here of average rates of advance in order to focus attention where it is needed.

The SDI power program should continue an aggressive, coordinated base technology program to parallel and complement its weapons platform/systems efforts. To enable the multidecades of advances needed for SDI power, program focus should be on areas such as:

- high-temperature materials for nuclear reactors and power generation;
- high-temperature radiators;
- advanced, high-temperature instrumentation and reactor control;
- two-phase flow evaporation and condensation in reduced gravitational fields;
- electrical and thermal insulators;
- low-mass electrical conductors, including superconductors;
- thermal conductors;
- ferromagnetic and magnetic materials;
- survivable devices for switching, power conditioning, and generation;
- techniques for managing/containing high voltages, currents, and electrical and magnetic fields; and
- improvements in inverters, which are not presently being developed for weapons power.

The committee recognizes a clear need to make progress in materials for increasing the efficiency and compactness of power components. There may also be benefit in coupling industry to university research groups via the SDIO directorates responsible for basic research (DOD category 6.1) and technology base development in the power area. As an example, mass reduction in high-power thyratrons could be substantial if the ceramic insulators could either be

TABLE 5-1 Power System Components

Element	Factor	Current Status	SDI Near-Term Needs (1995)	Observed Evolutionary Advance Rate	Projected Date to Meet SDI Needs, Evolutionary Rate (year)	Representative Current Advance Rate (per decade)	Data to Near-Term Needs With Present Funding (year)
Capacitor	Energy density and scaling	Burst 300 J/kg	2,000 J/kg	x 2.2 per decade	2011	x 18	1993
	Energy density and life	Continuous 3 J/kg	50 J/kg	x 2.2 per decade	2022	x 18	1997
Magnetic modulators	Rep-rate power duty	1 MW short burst	10 MW	x 2 per decade	2019	x 1,600	1990

Resistor	Field and voltage stress levels	10[illegible] V/m, DC	10[illegible] V/m	x 1.[illegible] per decade	[illegible]	x [illegible]	[illegible]
Semiconductor diodes	continuous current	1 kA, 1 kV	10 kA	x 78 per decade (funded)	1997	x 78	1997
Thyratrons	Energy density	70 kW/kg	0.5 MW/kg	x 2 per decade	2015	x 10	1995
Transformers, inverter	Power and energy	30 kW 20 kHz AC	1 MW power level	x 2.7 per decade (in power-level density)	2020	x 80	1995
Transistors	Voltage	250 V	5 x 10^3	x 75 per decade	1994	x 75	1994
	Current	10^3 A	10^4 A	(funded)	1992	x 75	1992

SOURCE: Based upon a briefing by W. J. Sarjeant in June 1985 to the Defense Technologies Study Team on Space Power Technology, Arlington, Virginia.

eliminated or made of less dense insulating materials capable of high-temperature operation.

Progress toward advances in the state of the art of components used in power-conditioning and pulsed-power systems could be effectively achieved through initiatives anchored in materials technologies. In view of the major successes achieved in applying basic science to materials programs in high-energy-density capacitors, similar approaches should be applied to other areas of power development. This committee recommends a development strategy of this nature, pursued aggressively and funded adequately, to develop scalable power technology, particularly if success would enable selection of one weapon system over a less desirable one by removing power considerations as the principal constraint.

The following four areas of development form an integrated program ensemble in both prime power and power-conditioning technology:

- Technology feasibility projects to demonstrate that a required capability is possible.
- Scaled experiments to give high confidence in the ability to design a full-size system.
- Limited near-full-scale demonstrations of advanced-development models, for technology validation and to clarify integration and compatibility problems associated with production devices.
- A continuous effort to understand fundamental mechanisms as applied to component technology feasibility and scalability.

In summary, development of subscale (i.e., at about 10 percent of full power level), scalable, high-performance power components and associated technology to provide a broad range of system options is a prudent investment strategy. Emphasis on component development for generating, conditioning, and transmitting electrical power is required. *The issue of high-temperature superconductivity as it affects scaling feasibility must be addressed.* Furthermore, the longer-term and nearer-term technology base development programs must be brought into balance. A technology-based-option investment strategy for the longer-term options in SDI is needed by periodically targeting superior technologies among existing candidates as a means of achieving future needs through down-selection. Such an increased emphasis is needed on the technology base for space power system components, as the existing base is grossly inadequate to meet the mission challenge.

MATERIALS ADVANCES REQUIRED FOR THE EVOLVING SPACE POWER TECHNOLOGIES

There are vast differences in the materials requirements for the range of space power cycles and power systems examined by this committee. These systems typically demand high temperatures (with little else specified) ranging from 1300°K to 2500°K. Although the lower temperatures in this range can be met in reasonable time and at reasonable cost, the higher ones will necessitate the development of totally new or different materials, requiring a dedicated effort in order to achieve success in some "short-term" period such as 10 years. The development of SDI space power component technologies will require significant advances of materials technology in the following areas of magnetic materials, insulators, and the development of high-temperature structural materials.

Magnetic Materials

Magnetic materials are important for induction accelerators, low-mass, high-frequency inverters, and so on. However, data are currently being obtained that indicate that FeNdB magnets can be fabricated for a variety of magnetic applications with outstanding results. Metallic glasses of selected compositions are soft magnetic materials. Being free of grains, grain boundaries, and secondary phases, these materials can be used for making soft magnetic alloys that are entirely free of orientation effects.

Apparently Metglass^R has not yet approached the potential desirable properties achievable in magnetic materials by applying rapid quenching techniques to create new alloys. During the past year, General Electric—using Allied Signal Company Metglass^R compositions—built and tested a large number of commercial AC power transformers that exhibited the outstanding performance previously predicted.

Insulators

Newly developed products far superior to classic baked clay ceramics are available for making feedthroughs, standoffs, interfaces, and other insulators. Numerous new classes of polymers and ceramics, processing techniques, and forming techniques can now offer major improvements in insulators. Such improvements include high-strength materials that can be used at high and low temperatures and that can produce intricate shapes.

High-Temperature Structural Materials

Because materials are almost always—and properly—viewed as design limiters, support for the development of advanced materials has received reasonable backing since the early 1950s. Unfortunately, performance specifications all too frequently come fairly late in systems development programs. Furthermore, almost every new application unfortunately requires new or different combinations of properties and performance: temperature, time at temperature, permissible deformation, structural stability (i.e., changes of properties under operating conditions), surface degradation, joining problems, and so on. For new or different applications, these requirements emphasize the need to define a proposed system so that materials can be tailored to such needs. It is rare that the more critical materials can be obtained "off the shelf."

For SDI power systems, radiation hardening is a requirement for power semiconductor switches and other electrical components. There are significant opportunities for exploiting new materials such as gallium arsenide and silicon carbide for this purpose.

Before the use of ceramic materials or carbon–carbon composites for rotating blades—or the use of filament-reinforced ceramics for temperatures between 1200°K and 1500°K—can be seriously proposed, considerable time will be needed to develop and test such materials for use in a specific power system. This is because only limited data are available on long-term performance in highly cyclical temperature and stress systems. A few such systems are making excellent progress, but results for these applications are emerging slowly, hence careful development of these materials for meeting specific needs will continue to be required.

The preferred cycles and systems must be selected, and all operating conditions must be integrated. Such integration will permit selection of the alloy systems, if not of the alloys, for preliminary consideration and planning for alloy modification. Thus the committee notes the following three partial bases for arriving at its Conclusion 5 and Recommendation 5 stated below.

1. Selection of operating temperatures up to about 1500°C (1773°K) (National Research Council, 1988) may permit preliminary

selection of materials already in existence for specified life cycles and environments. Usually, and fairly obviously, the lower the planned operating temperatures, the greater is the number of available applicable alloys. The refractory metal alloys are reasonably well known and perform well at the right temperatures and atmospheric pressures, but must be carefully selected for ductility.

2. For temperature applications above about 1100°C–1200°C (1373°K–1473°K) regardless of alloy type (metallic base, ceramic base, carbon base), only limited data are available for lifetimes in excess of 100 h—or even for lifetimes in excess of only a few hours—although there are important exceptions. Obviously, low-mass structures should be emphasized.

3. Coatings may be required for advanced materials operating at high temperatures for significant periods. This area has received very little funding, yet it is critical for the selection of appropriate materials.

CONCLUSION AND RECOMMENDATION

Based on the discussion in this chapter, the committee arrived at the following conclusion and recommendation.

Conclusion 5: Major advances in materials, components, and power system technology will be determining factors in making SDI space power systems viable. Achieving such advances will require skills, time, money, and significant technological innovation. The development of adequate power supplies may well pace the entire SDI program.

Recommendation 5: Make additional and effective investments now in technology and demonstrations leading to advanced components, including but not limited to:

- **thermal management, including radiators;**
- **materials—structural, thermal, environmental, and superconducting;**
- **electrical generation, conditioning, switching, transmission, and storage; and**
- **long-term cryostorage of H_2 and O_2.**

Advances in these areas will reduce power system mass and environmental impacts, improve power system reliability, and, in the long term, reduce life-cycle power system cost.

6
Commentaries on the SDI Power Program

COMMENTARY ON SDI SPACECRAFT SYSTEM NEEDS AND THEIR IMPACTS ON THE SPACE POWER SYSTEM

There are several spacecraft-level SDI system needs that could well affect the space power system but could not be assessed by the committee because of the limited scope of available studies. This recognition underlies the committee's recommendation for representative all-up preliminary spacecraft designs (Recommendation 1). The following list of system needs that require better definition is meant to be representative but not all-inclusive:

- vehicle maximum slewing rates, presumably established by needs for retargeting;
- verification of spacecraft operational readiness;
- vehicle-maneuvering requirements, including replenishment of orbital-drag losses as well as needs for evasive actions;
- elimination of torques produced by spacecraft interaction with the earth's magnetic field; and
- modification of spacecraft thermal and radar signatures.

COMMENTARY ON SDI PROGRAM ISSUES

In this section, the committee has highlighted the programmatic concerns it believes to be most important among the recurring space power issues confronting SDIO management. These issues are

a. Balancing investment of resources between the near term and the long term.

b. Coordinating the investment in basic technologies and components to produce timely results and to emphasize high-leverage items.

c. Integration of SDI power supply systems with overall SDI systems.

The dilemma posed in (a) above is apparent in both large and small examples. In Chapter 3, the point was made regarding nuclear power that failing to initiate and carry out the development of a multimegawatt power source would limit SDI options for very-high-power electrically energized systems.

Not so immediately obvious is the potential referred to in (b) for crippling the future by failing to develop critical advanced technology components needed by SDI, such as resistors, insulating materials, high-temperature structural materials, thyratrons, transformers, and so on. The unstimulated rate of improvement is, in many instances, not rapid enough. Stimulating power-system-component development will probably have beneficial impacts on many technical activities in addition to SDI, a factor in motivating Recommendation 8.

Issue (c) expresses the committee's concern that the overall space platform will impose demands on the power supply system not immediately apparent when examining the space power supply system in isolation. Active Program Management and Integration at the vehicle level can address the issues cited, and should examine with skepticism all estimates of development times and costs. The length of time required to design, develop, and qualify new power sources—especially nuclear ones—must be a key consideration, a factor motivating Recommendations 4 and 8.

With this introduction, a more detailed commentary on SDIO budget allocation and strategy follows.

REVIEW OF THE SDI SPACE POWER PROGRAM

The SDIO fiscal year (FY) 1987 and 1988 budgets, future projected budgets, and interviews with staff in the SDIO Power Program Office were used as data to deduce the current SDIO investment strategy for its space power program. This present strategy includes five elements:

1. Provide power for initial deployment based on chemical rockets and passive sensor platforms.
2. Conduct technical research and development for providing power to directed-energy weapons for discriminating decoys interactively and for killing boosters, postboost vehicles, and warheads.
3. Demonstrate enabling long-term technology. (An "enabling" technology is one that satisfies an applications requirement.)
4. Formulate future requirements to guide development.
5. Provide power for near-term SDI experiments and tests.

Providing for near-term requirements (item 1) means enhancing survivable solar power in FY 1988 and beginning an exploratory examination of a small nuclear reactor power system at approximately 1 percent of the FY 1988 SDIO space power budget. This shift is a direct response to near-term requirements, and shows technically agile and responsive program management.

The high priority SDIO has assigned for conducting technology development for powering directed-energy systems (item 2) is necessary to advance the rate of progress in pulsed-power development and to avert a century of development to attain gigawatt power levels for space missions. Substantial R&D is required to reduce that time from a century to 10–20 years. The emphasis on technology development is appropriately placed, and can provide for long-term SDI needs.

Demonstrating enabling long-term technology (item 3) has led to continuing support for the SP-100, which was 48 percent of the FY 1987 budget and will be 30 percent of the FY 1988 budget. The SP-100 project could provide enabling technology that is suited for powering many missions in space, on the moon, and for exploration of the solar system. History has shown that it takes longer to develop a nuclear reactor system, such as SP-100, takes longer than to develop a space mission. Hence today's civil and military space project managers cannot include any nuclear reactor space power system—or any other system—in their mission planning until that system has been developed and tested. This dilemma is frequently referred to as the "chicken-and-egg syndrome." Consequently, support of the SP-100 program by the SDIO Power Program Office—in the absence of a commitment for a specific space mission—shows a farsighted perspective, and this committee strongly endorses that strategy. An operational SP-100 space power system would be a major product of—and a major benefit to—the SDI effort.

The power portion of the SDI program is currently in a component development and technology phase. Until SDI weapons systems have been more clearly defined, more aggressive advanced development and engineering of SDI power systems would—in most instances—be premature. Based upon the personal experience of several of its members, this committee finds that investment in large, narrow demonstrations is generally wasted if it is accomplished more than 6 to 10 years before deployment. Accordingly, this committee strongly encourages the development of generic, scalable space power technologies for the future, but believes that advanced development of specific space power systems should await improved specification of SDI weapons. The SP-100 is an exception, and is justified by its having broad applicability to civil and military space power missions and by its potential for providing experience on how to integrate pulsed-power systems in space.

Development of technology for superconducting magnetic energy storage (SMES) is a response to the power requirement for the ground-based (free-electron) laser, and is also motivated by a potential spinoff to load leveling of commercial power. Enabling SMES technology could also result in substantial overall economic benefits to this country, in addition to technical benefits to SDI.

The pace of developing repetitive, high-powered, pulsed-power systems needs to be accelerated as much as possible. Over the last several decades, the power available from continuously operated repetitive pulsed-power systems has increased every decade by a factor of approximately three. In the early 1980s, average power levels of accessible terrestrial power-conditioning technology were in the neighborhood of 1 MWe. Given that technologies can usually be adapted to space, one could envision a multimegawatt capability in the early 1990s; however, pulsed-power systems for hundreds of megawatts to several gigawatts may be required for some SDI weapons platforms. Their development could require 40 to 60 years at the demonstrated historical extrapolation rate, making them unavailable until about 2030 to 2050. It may be possibile to dramatically accelerate this rate in some areas by developing new materials and by stimulating innovative technological approaches to space power system applications. Such acceleration has already been achieved in the capacitor development program, and similar approaches are needed in other power areas.

The current emphasis on multimegawatt space power systems seems to be appropriate, but the goals of the SDI space power program require several hundreds of megawatts. The next round of power architecture studies and power system studies must address the need for aggressively advancing space power technologies to the multihundred-megawatt class.

Formulating future requirements (item 4) to guide development is an essential part of every technically challenging program. The three Space Power Architecture System (SPAS) studies (1988)—constituting about one percent of the SDIO power program budget in FY 1987—were an initial attempt to establish relative priorities of various candidate space power system concepts. However, during the course of the committee's study, the SPAS studies had still not been published. Because the architects studied a wide variety of power systems and employed differing assumptions, the committee found it exceedingly difficult to make direct comparisons among the results of the three studies. Nevertheless, the requirements definition for those studies identified some key total-system issues, such as the effluent-tolerability issue.

There is substantial interaction within SDIO between its Power Program Office and its programs dealing with kinetic energy weapons, directed-energy weapons, and sensors, in order to provide qualitative and quantitative guidance for emphasis within the space power program. The committee finds that the SDIO Power Program Office is being responsive to the changing needs of the other SDIO directorates by shaping its program accordingly.

Providing power for near-term SDI experiments and tests (item 5) is an ancillary strategy. Superconducting Magnetic Energy Storage would support near-term testing of a ground-based free-electron laser (FEL) if an SMES system were sited at White Sands, and would provide technology for future FEL tests if one were located elsewhere. Magnetohydrodynamic (MHD) power generation was directed to near-term SDI power requirements because the resulting effluents may make the use of MHD in space unattractive in the long-term.

In addition to these basic strategies, the committee notes that SDI requirements often change. Various weapons systems have waxed and waned in popularity, resulting in shifting SDI power system requirements. Such shifts in requirements will continue as the major technical aspects of each weapon concept become better defined. The pursuit of long-term power system demonstrations should be buffered from such short-term fluctuations because significant resources and continuity would otherwise be wasted in starting and stopping demonstration projects. This is a factor leading to Recommendation 8.

The space power technology development program, which can lead to large advances in capability, should also be protected from shifting requirements, because it has large potential long-term benefits for a relatively low annual investment. Thus this program requires consistent year-to-year support during a protracted development cycle.

Program management issues were also examined during the study. The committee found that the relative power program funding allocations among the five major program investment strategies being pursued have been reasonable. The SDIO power program has been responsive to the needs of SDI users, both to share in the long-term benefits of space power development and to transfer technology to them in order to achieve near-term gains. It will be very difficult to supply space power with assured reliability, provide it in a form matching user needs, and make it available promptly on demand during periods when it is needed. This difficulty needs to be appreciated within the various SDIO directorates.

A more detailed examination of the SDIO space power program investment strategy for the FY 1987 and FY 1988 budgets follows.

COMMENTARY ON THE SDI SPACE POWER INVESTMENT STRATEGY

The SDIO space power program budgets for FY 1987 and FY 1988 were reviewed to deduce the current SDIO investment strategy for space power. The raw data are presented in Figures 6-1 through 6-4.

The total space power program budget increased from $80 million in FY 1987 to $95 million in FY 1988, and is distributed as shown in Figure 6-1. Funding for integrated demonstrations made up almost half the total budget in FY 1987, and a lesser fraction in FY 1988. Research on components received roughly the same percentage, about 20 percent, in both fiscal years. Funding for research on generation technologies increased substantially in FY 1988 as funding for power systems studies—including programmatic support—decreased. Although the committee regarded the increase in funding in the area of generation technologies as well warranted,

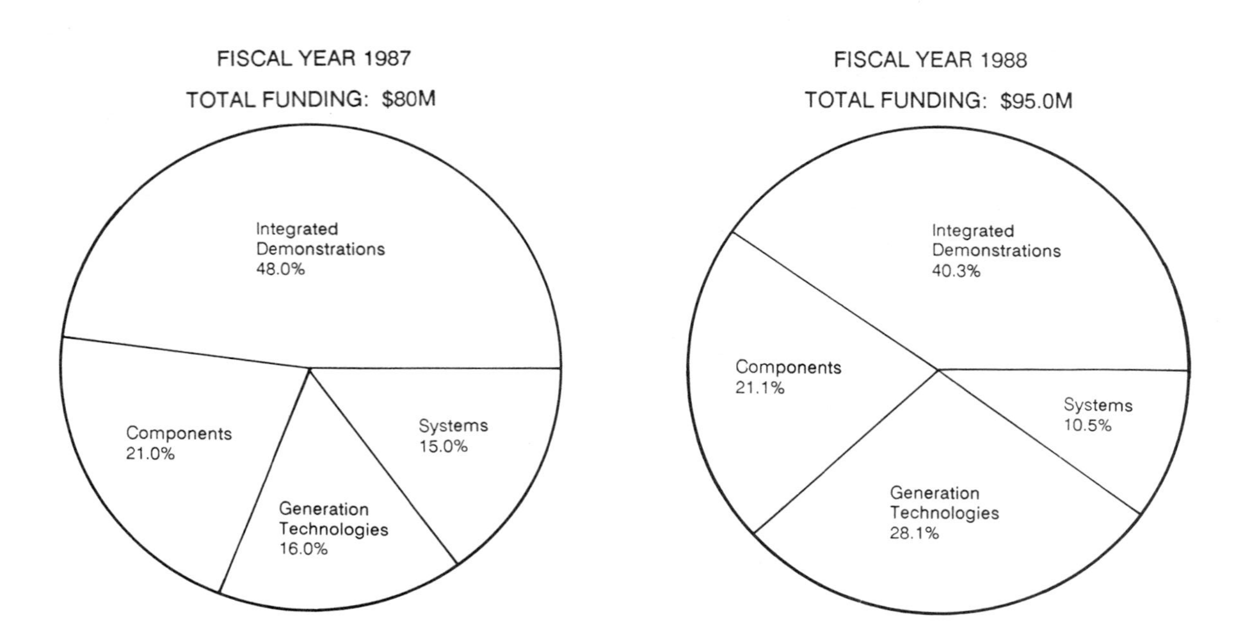

FIGURE 6-1 Budgetary breakdown of SDI R&D on space power for fiscal years 1987 and 1988.

additional studies of the space power subsystem as an integral part of a space weapons system are a high-priority need. This recognition is reflected in Recommendation 1.

Development of SP-100 is a joint program of the DOD (through SDIO), DOE, and NASA. Joint funding of SP-100 by the three agencies totalled about $60 million in FY 1987 and $99 million in FY 1988. The SP-100 was essentially the SDIO Power Program's only integrated demonstration during FY 1987, and received $38 million of SDI funding. In FY 1988 the SDIO portion of SP-100 funding was $25 million, and a new SMES program is funded at $13 million to provide a near-term capability. Cost sharing of the SP-100 program by other agencies permitted the total SP-100 program to grow, even though the SDIO funding component declined. Since all orbiting platforms will require housekeeping power, and the versatile SP-100 system has been designed to provide the range of powers needed for that purpose, the committee regards demonstration of a space nuclear reactor power system as one way of implementing Recommendation 4, since it would provide valuable experience in how to integrate various components into a space power system.

The committee cautions, however, that growth in SP-100 funding should not consume the remainder of federal funds allocated for SDI space power development. Thus, the committee reluctantly recommends slowing down joint SDI/DOE/NASA development of SP-100—but not below a "critical" rate—if that step is essential to preserve the SDI power technology program at a viable level. If a substantial portion of SP-100 funding can be provided by the DOE and NASA, SP-100 development can proceed without consuming the entire SDI power program budget. The committee strongly encourages SDIO to pursue these project partnerships aggressively to avoid erosion of its power technology development program. The SDIO strategy for FY 1988 maintains a strong power technology development program along with SP-100. Although the committee favors SDI funding of SMES, it is concerned that this not be accomplished at the expense of SP-100 development.

There are five multimegawatt power generation technologies being pursued with the $13 million in FY 1987 and $26.7 million in FY 1988 devoted to technology development. The subdivision of that sum is shown in Figure 6-2. The multimegawatt nuclear power program, in collaboration with the DOE, is the largest program, but has a decreasing budgetary share from SDIO in FY 1988. The electrochemical technologies enjoyed reasonable growth. The major

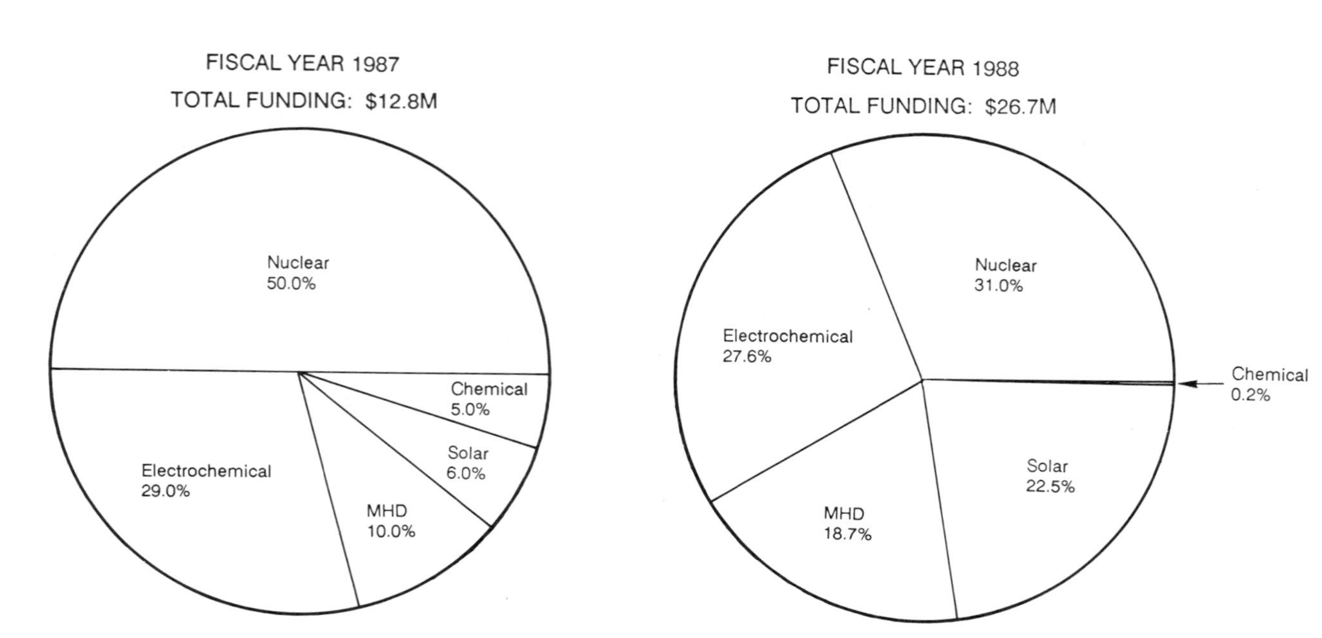

FIGURE 6-2 Funding of the five multimegawatt power generation technologies being pursued. MHD = magnetohydrodynamics.

growth in R&D on survivable solar power technology reflects near-term needs. Because of its potential as a compact multimegawatt power source for weapons (with only the hydrogen used for thermal management as an effluent), the committee understands why nuclear power was a major portion of the program. The electrochemical program may be justified by the potential of fuel cells to provide rechargeable power for system start-up at battle time. The fuel cells can be arranged in modular series parallel combinations to minimize power conditioning. Battery research is being funded aggressively for a short time to stimulate rapid maturation of that technology.

The MHD program, which grew substantially from FY 1987 to FY 1988, appears to command a disproportionate amount of funding. Such a spending profile might be justifiable to satisfy a near-term program need for powering ground tests if the project would make available a power facility for ground tests at lower overall cost than that of more conventional alternative technologies *and* if that facility could be available with high reliability. The large investment for a short time appears to be regarded by SDIO as the most cost-effective way to produce this near-term power source. The committee regards the effluents associated with an MHD space power system—or with other open-cycle power systems—as potentially incompatible with long-term SDIO needs for space power.

The major decrease in funds for chemical technology is justified because it is an off-the-shelf technology. Should this option be selected for deployment in space, its development, integration, and space qualification will require a major program.

Component development is being broadly pursued in the program, receiving approximately $14 million of the FY 1987 funds, and $20.8 million in FY 1988, divided as shown in Figure 6-3. The large portion going to radio frequency power sources is a program response to the current emphasis on the neutral-particle beam and the free-electron laser.

In FY 1987 the second largest budgetary portion, 19 percent, was for R&D on inductive storage and switching, a program element that is a holdover from prior commitments when electromagnetic launchers were given high priority in SDI. Funding trends for FY 1988 reduced this commitment appropriately.

Funding for research on rotating machinery technology for power conditioning grew from FY 1987 to FY 1988 because that technology permits reasonable pulse compression to the millisecond range. The committee understands that current-collection technology is a major

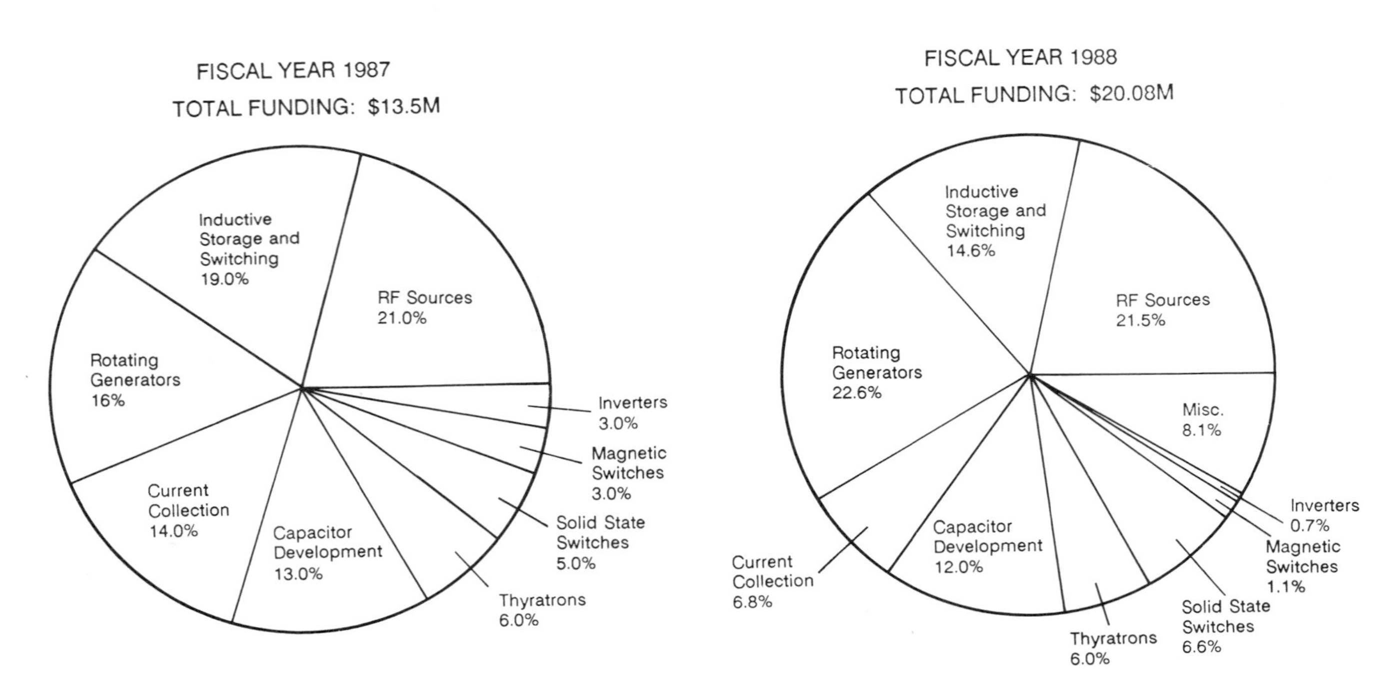

FIGURE 6-3 Funding breakdown for component development. RF = radio frequency.

limitation and, as such, it represented 14 percent of the total budget in FY 1987. The effort on current collection associated with large inductive storage and switching is appropriately reduced in FY 1988. The current-collection work for general power conditioning may be continued if improvements in power conditioning capabilities are required.

The capacitor development program, at 12 percent to 13 percent of total funding, has been an outstanding success. The application of science to enable creating materials for high-strength, high-energy-density capacitors in the current SDI program has been quite impressive. This successful and exemplary development of high-energy-density capacitors and the demonstration of their reliability are likely to have very broad applications in SDI, tactical defense, and a large number of significant civil and military programs.

Budget allocations for exploration and development of thyratrons, solid-state switches, and magnetic switches are reasonable. It would be highly desirable for science to be brought to bear in catalyzing the development of these devices with the goal of successes similar to those achieved in capacitor development.

In contrast to development of the above technologies, development of inverters, at 3 percent in FY 1987 and 0.7 percent in FY 1988, seems grossly underfunded, and the funding trend is in the wrong direction. In particular, there is currently no development of technology for the high-duty-factor solid-state switches, transformers, and capacitors required for multimegawatt inverters. This deficiency should be corrected by implementing an aggressive inverter program, including the component development needed for those inverters.

System and system technologies have three explicit subdivisions totaling approximately $12 million. As shown in Figure 6-4, most of these funds are to define requirements, conduct surveys, fund program management of the SDIO Power Program Office, obtain program advice, and support the independent evaluation group (IEG).

Studies on thermal management are broadly applicable to all candidate power system concepts, and are strongly supported by the committee. Although survivability is not explicitly called out in any of the work package directives, it appears to be included in the work of the IEG.

From FY 1987 to FY 1988, the 38 percent reduction of funding for resolving environmental concerns would be of concern if it were to result in delaying resolution of the effluent issue. The capability of SDI weapons platforms to function when immersed in their own

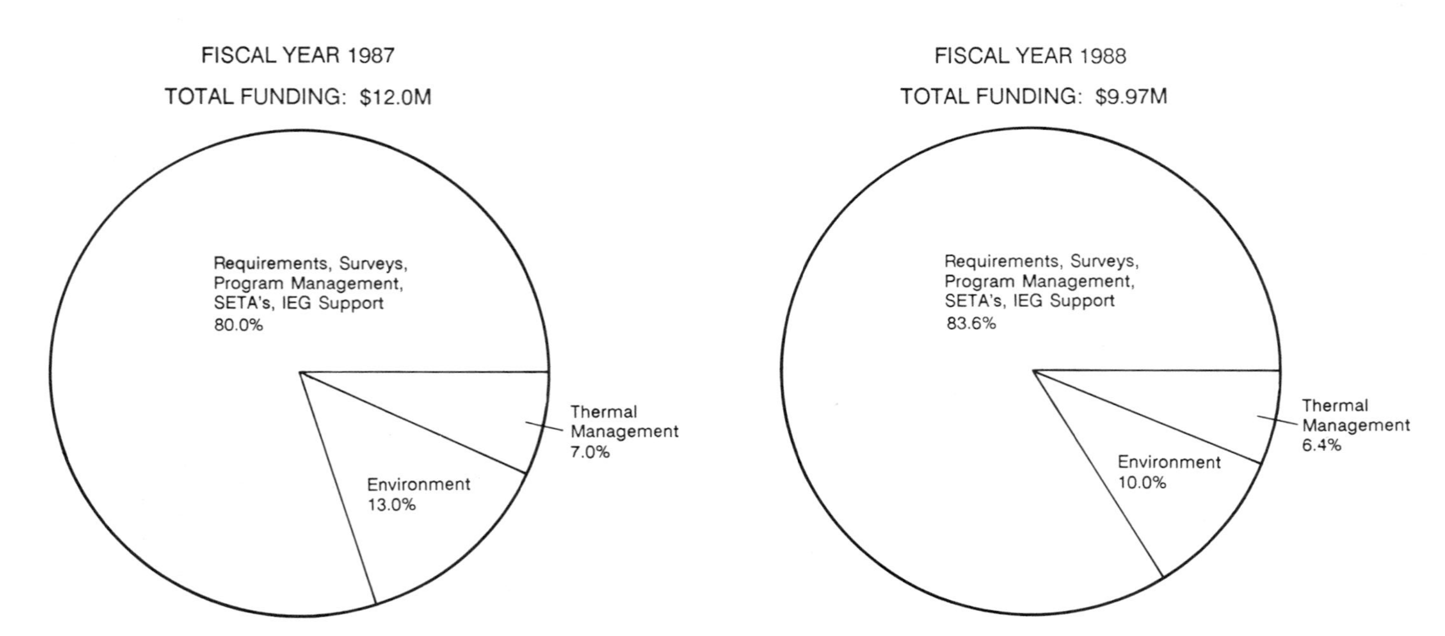

FIGURE 6-4 Funding distribution for system and system technologies. IEG = Independent Evaluation Group, SETA = Scientific, Engineering, and Technical Assistance.

effluents will have a strong impact on the selection of candidate multimegawatt technologies. If no effluent can be tolerated, the extra mass of closed-cycle systems will be costly. If hydrogen is the only acceptable effluent, then space power systems would be restricted to that effluent, and storage would be required aboard the spacecraft of chemical combustion products or of effluents from nuclear open-cycle power systems. As stated in Conclusion 3 and Recommendation 2 (Chapter 3), the committee recommends a concerted study of the effluent issue to discriminate among power options before selecting the most promising approaches. Of course, if some weapons platforms can tolerate substantial effluent, the improved power-per-unit mass from the open-cycle systems and the reduced reliance on the nuclear option—with its associated environmental and survivability problems—may justify developing several multimegawatt power options.

FINDING, CONCLUSION, AND RECOMMENDATIONS

Based on the discussion in this chapter, the committee arrived at the following finding, conclusion, and recommendations:

Finding 7: The present overall rate of progress in improving the capability of space power conversion and power-conditioning components appears inadequate to meet SDI schedules or NASA needs beyond the Space Station.

Conclusion 6: Refocusing SDIO resources toward near-term weapons systems demonstrations could delay development of advanced power technology, and thereby seriously jeopardize meeting long-term space power program objectives.

Recommendation 3b: The SP-100 nuclear power system is applicable both to SDI requirements and to other civil and military space missions. Therefore, SP-100 development should be completed, following critical reviews of SP-100 performance goals, design, and design margins.

Recommendation 3c: SDI burst-mode requirements exceed by one or more orders of magnitude the maximum power output of the SP-100. Therefore, both the nuclear and nonnuclear SDI

multimegawatt programs should be pursued. Hardware development should be coordinated with the results of implementing Recommendation 5.

Recommendation 4: Consider deploying the SP-100 or a chemical power system on an unmanned orbital platform at an early date. Such an orbital "wall socket" could power a number of scientific and engineering experiments. It would concurrently provide experience relevant to practical operation of a space power system similar to systems that might be required by the SDI alert and burst modes.

Recommendation 8: To further U.S. capabilities and progress in civil as well as military applications of power technology, both on the ground and in space, and to maintain a rate of progress in advanced technologies adequate to satisfy national needs for space power, plan and implement a focused federal program to develop the requisite space power technologies and systems. This program—based on a multiyear federal commitment—should be at least as large as the present combined NASA, DOD (including SDIO), and DOE space power programs, independent of the extent to which SDI itself is funded.

References

Billman, Fred. 1987. High Purity Aluminum Conductors. Electric Generator Developers Conference. (Alexandria, Virginia, January 21–22, 1987.) U.S. Army Belvoir RD&E Center Report AD-B112-729.

Brown, William C. 1987. LEO to GEO Transportation System Combining Electric Propulsion with Beamed Microwave Power from Earth. Proceedings, AAS Goddard Memorial Symposium (NASA Goddard Space Flight Center, March 18–19, 1987).

Buckman, R. W., Jr., and R. R. Begley. 1969. Development of High-Strength Tantalum Base Alloys. Pp. 19–37 in *Recent Advances in Refractory Alloys for Space Power Systems*. NASA report SP-245. June.

Buden, David. 1981. The acceptability of reactors in space. Report LA-8724-MS, Los Alamos National Laboratory (UC-33). April.

Colladay, Raymond S., and Edward A. Gabris. 1988. Presentation to Space Nuclear Power Systems Symposium. Albuquerque, New Mexico. January.

Cooper, R., and J. Horak. 1984. In Proceedings of the Symposium on Refractory Alloy Technology for Space Nuclear Power Applications (Oak Ridge, Tennessee, August 10–11, 1983), R. H. Cooper, Jr. and E. E. Hoffman, eds. Report CONF-8308130 (also, Report DE 84-001745). January.

Cropp, Louis. 1988. Sandía National Laboratories, IEG Field Support Team, personal communication.

DeVan, J. H., and E. L. Long. 1975. Evaluation of T-111 Forced-Convection Loop Tested with Lithium at 1370°C. NASA report CR-134745. (Also published as Oak Ridge National Laboratory report ORNL TM-4775.)

DeVan, J. H., J. R. DiStefano, and E. E. Hoffman. 1984. Compatability of Refractory Alloys with Space Reactor System Coolants and Working Fluids. Pp. 34–85 in *Refractory Alloy Technology for Space Nuclear Power Applications*, R. H. Cooper, Jr., and E. E. Hoffman, eds., DOE CONF-8301B130.

Eckels, Phillip. 1987. High Purity Aluminum. Electric Generator Developers Conference (Alexandria, Virginia, January 21–22, 1987). U.S. Army Belvoir RD&E Center Report AD-B112-729.

El-Genk, M. S., A. G. Parlos, J. M. McGhee, S. Lapin, D. Buden, and J. Mims. 1987. Pellet Bed Reactor Design for Space Power. Proceedings of 22nd Intersociety Energy Conversion Engineering Conference (Philadelphia, August 10–14, 1987). American Institute of Aeronautics and Astronautics, Washington, D.C.

English, Robert E. 1978. Alternative Power-Generating Systems. Conference on Future Orbital Power Systems Technology Requirements. NASA CP-2058.

English, Robert E. 1987. Speculations on Future Opportunities to Evolve Brayton Power Plants Aboard the Space Station. Proceedings, 4th Symposium on Space Nuclear Power Systems, Albuquerque, New Mexico. January.

Gas Turbine World. 1987. Handbook. 10:2–6.

Gregorwich, W. S. 1987. Microwave Power Beaming from Earth-to-Space. Lockheed Research Laboratories, Palo Alto, California.

Hoffert, M. I., B. Heilweil, G. Miller, W. Ziegler, and M. Kadiramangalam. 1987. Earth-to-Satellite Microwave Beams: A Non-Nuclear Approach to SDI Space Power. Department of Applied Science, New York University.

Johnson, Colonel Joseph R. 1988. Testimony on Space Nuclear Reactor Power Systems, presented to the U.S. House of Representatives Committee on Science, Space, and Technology, March 16.

Johnson, Nicholas L. 1986. Nuclear Power Supplies in Orbit. Space Policy, pp. 227–228. August.

Johnstone, John A. 1988. Argonne National Laboratory, personal communication.

Klopp, William D., Robert H. Titran, and Keith D. Sheffler. 1980. Long-Term Creep Behavior of the Tantalum Alloy Astar 811C. NASA TP 1691. September.

Kulcinski, G. L., and H. H. Schmitt. 1987. The Moon: An Abundant Source of Clean and Safe Fusion Fuel for the 21st Century. UWFDM-730, Fusion Technology Institute, University of Wisconsin, Madison. August.

Lundberg, Lynn B. 1985. Refractory metals in space nuclear power. Journal of Metals, April:44–47.

Mankins, J., J. Olivieri, and A. Hepenstal. 1987. Preliminary Survey of 21st Century Civil Mission Applications of Space Nuclear Power. JPL D-3547. Pasadena, California: Jet Propulsion Laboratory. March.

Materials Research Society Conference. 1987. Georgia Institute of Technology. December.

Murphy, G. B., S. D. Shawhan, and J. S. Pickett. 1983. "Perturbations to the plasma environment induced by the Orbiter's maneuvering thrusters," American Institute of Aeronautics and Astronautics (AIAA) Paper 83-2599, New York and Washington. October.

NASA Internal Study. 1988.

National Research Council. 1983. Committee on Advanced Nuclear Systems, "Advanced Nuclear Systems for Portable Power in Space," Washington, D.C.: National Academy Press.

National Research Council. 1988. Report of a study by the Committee on High-Temperature Materials for Advanced Technological Applications. Washington, D.C.: National Academy Press.

Pickett, J. S., et al. 1985. "Effects of chemical releases by the STS 3 Orbiter on the ionosphere," Journal of Geophysical Research, 90: 3487–3497.

Ride, Sally K. 1987 Leadership and America's Future in Space. A report to the Administrator of NASA. Washington, D.C.: NASA. August.

Rohwein, G. J., and W. J. Sarjeant. 1983. Critical Issues in Electrical Energy Storage and Transfer. Applied Physics Communications, 3:169–210.

Rosenblum, L., D. R. Englund, Jr., R. W. Hall, T. A. Moss, and C. Scheuermann. 1966. Potassium Rankine System Materials Technology. Pp. 169–199 in *Proceedings, Space Power Systems Advanced Technology Conference.* NASA report SP-131.

Sarjeant, W. J. 1985. Briefing to the Defense Technology Study Team on Space Power Technology, Arlington, Virginia. June.

Seaton, Michael K. 1985. A summary of the state of the art of nuclear safety of space nuclear reactors. Air Force Weapons Laboratory Report AFWL-TR-84-144, Dayton, Ohio. September.

Shawhan, S. D., and G. B. Murphy. 1983. "Plasma diagnostics package assessment of the STS-3 Orbiter environment and systems for science," AIAA Paper 83-0253, New York and Washington, January.

Space Power Architecture System (SPAS) studies. 1988. Technical evaluations by the Field Support Team of the SDIO Power Program Office's Independent Evaluation Group of draft final reports by three contractors: General Electric, Martin Marietta, and TRW.

Stephens, Joseph R., Donald W. Petrasek, and Robert H. Titran. 1988. Refractory Metal Alloys and Composites for Space Power Systems. NASA report TM 100946.

Strategic Defense Initiative Organization (SDIO). 1986. SDI Survivability Technology Hardening Goal. WPD-L-004: Space Science Passive Survivability Technology FY87. October.

Symposium on Space Nuclear Power Systems. 1988 (and previous years). University of New Mexico. Proceedings, Mohamed El-Genk, ed. Albuquerque, New Mexico.

USAF/DOE Evaluation Panel for Small Space Reactor Systems. 1988. Space Reactor Power Systems for 5 to 40 Kilowatts. Air Force Space Technology Center and U.S. Department of Energy. March.

APPENDIX

A
Glossary of Abbreviations

AEC	Atomic Energy Commission
ARDEC-KAMAN	Army Research and Development Center-Kaman Company
BSTS	Boost Surveillance and Tracking System
BTI	Balanced Technology Initiative
C^3	Command, Communications, and Control
CARDS	Concept and Requirements Definition Study
CPB	Charged particle beam
DARPA	Defense Advanced Research Projects Agency
DEW	Directed-energy weapon
DIPS	Dynamic Isotope Power Sources
DNA	Defense Nuclear Agency
DOD	Department of Defense
FEL	Free-Electron Laser
FES	Flash Evaporator System
GEO	Geosynchronous orbit
GES	Ground Engineering System
IEG	Independent Evaluation Group
KEW	Kinetic Energy Weapon
LEO	Low earth orbit
LIA	Linear Induction Accelerator
MHD	Magnetohydrodynamics
MMH	Monomethyl Hydrazine

MMW	Multimegawatt
MMWSS	Multimegawatt Steady State
NASA	National Aeronautics and Space Administration
NPB	Neutral-particle beam
ODS	Oxide-dispersion-stabilized
OMS	Orbital Maneuvering System
PDP	Plasma diagnostics package
PM&I	Program Management and Integration
PRCS	Primary Reaction Control System
RCS	Reaction Control System
RF	Radio frequency
RTG	Radioisotope Thermoelectric Generator
SATKA	Surveillance, Acquisition, Tracking, and Kill Assessment
SDI	Strategic Defense Initiative
SDIO	Strategic Defense Initiative Organization
SETA	Scientific, Engineering, and Technical Assistance
SMES	Superconducting Magnetic Energy Storage
SNAP	Space Nuclear Reactor Program
SPAS	Space Power Architecture System
SPEAR	Space Plasma Experiments Aboard Rockets
TFE	Thermionic Fuel Element
USAF	U.S. Air Force

APPENDIX

B
Biographical Sketches

COMMITTEE MEMBERS

JOSEPH G. GAVIN, Jr. (*Chairman*)
Senior Management Consultant, Grumman Corporation

JOSEPH GAVIN retired as President of Grumman Corporation in September 1985 and is presently a Senior Management Consultant with Grumman. He received B.S. and M.S. degrees in aeronautical engineering from the Massachusetts Institute of Technology (MIT). Mr. Gavin has participated in aerospace design, engineering, and technology research and development as chief experimental project engineer, chief missile and space engineer, vice president, president, and chief operating officer and director at Grumman during a 39-year career span. He was the vice president responsible for design, construction, and mission support of the Apollo Lunar Module. He concurrently served as chairman or member of various government and industry committees and boards. In 1971 Mr. Gavin received the Distinguished Public Service Medal from the National Aeronautics and Space Administration. He is a member of the National Academy of Engineering, and a fellow of the American Astronomical Society and the American Institute of Aeronautics and Astronautics.

Prior to becoming a part-time consultant with Grumman Corporation at the end of September 1985, while president and chief operating officer there, he sponsored the preparation of proposals responding to requests for proposal from the Strategic Defense Initiative Organization (SDIO). He did not administer, nor does he now relate to, any resulting contract. Grumman has several SDI contracts, relating to neutral-particle beam platforms, boost surveillance tracking systems, and space nuclear power. It is not among the top SDI contractors nationwide. Mr. Gavin is a member of the Executive Committee of the MIT Corporation, and is on the administrative policy board for MIT and MIT Lincoln Laboratory. MIT as an institution does not participate contractually in SDI research; several of its faculty members do. MIT Lincoln Laboratory is among the top SDI contractors nationwide. Mr. Gavin is an outside director of the Charles Stark Draper Laboratories, which specializes in providing the U.S. government with research and prototypes in the field of inertial guidance and control mechanisms.

TOMMY R. BURKES (*Member*)
Visiting Professor of Electrical Engineering, Texas Tech University

TOMMY BURKES is a Visiting Professor of Electrical Engineering at Texas Tech University, Director of Technical Programs in its Center for Advanced Research, and President of T. R. Burkes, Inc. He has B.S.E.E. and M.S.E.E. degrees from Texas Tech University and a Ph.D. degree from Texas A&M University. Dr. Burkes has held associate and assistant professorships at Texas Tech as well. The focus of his recent professional activity is in the areas of high-voltage/pulse power electronics for driving lasers, electron beams, and microwave generators for radars and accelerators, as well as research in high-voltage fast switching, energy storage, and other power system components. Dr. Burkes organized and chaired the First International Pulse Power Conference in 1976. He has served as session chairman at a variety of conferences, symposiums, and workshops, and is a member of various technical evaluation teams for the Air Force, Army, Navy, and Strategic Defense Initiative Organization. He belongs to numerous honor societies.

Professor Burkes is an occasional consultant, about five days per year over the past two years, for the SDIO Power Program Office via W. J. Schafer Associates, Inc. This activity relates to technical review of existing contracts. In February 1988 he was appointed to a Defense Nuclear Agency (DNA) technical advisory group relating to the Superconducting Magnetic Energy Storage (SMES) program of SDIO. In 1986 he led a U.S. Air Force study effort regarding the applicability of the electrical utility power grid to powering ground-based lasers. Professor Burkes' department at Texas Tech University has two contracts with SDIO. He is not a principal investigator on either project.

ROBERT E. ENGLISH (*Member*)
Distinguished Research Associate and Consultant, NASA Lewis Research Center

ROBERT ENGLISH is a Distinguished Research Associate and Consultant at NASA Lewis Research Center in the fields of space power generation and electric propulsion. He serves without compensation. He holds B.S. and M.S. degrees in mechanical engineering from the University of Minnesota. Mr. English joined the Lewis Research Center in 1944, where he performed research on turbojet engines, then on generating electricity in space using both solar and nuclear power. When NASA was formed in 1958, English's work on electric propulsion for manned exploration of Mars formed the basis for the Lewis research program on space power and electric propulsion. On retiring in 1980, Mr. English was Deputy Director of Energy Programs, following a period as Chief of the Space Power Division. NASA awarded him its Exceptional Service Medal in 1975 and again in 1984. Mr. English has served on numerous committees and panels, notably for the Atomic Energy Commission, the Council on Environmental Quality, the National Institutes of Health, and the President's Office of Energy Policy and Commission on the Accident at Three Mile Island.

NICHOLAS J. GRANT (*Member*)
Professor of Materials Science and Engineering, Massachusetts Institute of Technology

NICHOLAS GRANT is a Professor of Materials Science and Engineering at MIT. He has a B.S. degree in materials science from the Carnegie Institute of Technology and an Sc.D. degree in metallurgy from MIT. Dr. Grant has held a number of academic positions in the field of metallurgy and materials science. He is the recipient of several awards and has led government-sponsored research projects and technical advisory committees. Fluent in Russian, Dr. Grant was chairman of the U.S. side of the U.S.–USSR science technology agreement on electrometallurgy and materials for the State Department from 1977 to 1982. He is a member of seven national engineering and science scholastic honoraries, including the National Academy of Engineering, and has participated in many National Academy of Sciences/National Research Council study committees, including a 1983 study on space nuclear power.

Professor Grant is not engaged in research relating to SDI. He has conducted numerous research projects sponsored by the U.S. Department of Energy (DOE); the Defense Advanced Research Projects Agency (DARPA); the U.S. Army, U.S. Navy, and U.S. Air Force; the National Aeronautics and Space Administration (NASA); the Atomic Energy Commission (AEC); the Office of Naval Research (ONR); and private industry. That research emphasized high-temperature materials for turbines, jet engines, nuclear power systems, and high-temperature structures. Studies have included alloy design and development and manufacture, testing and evaluation over a wide range of conditions.

GERALD L. KULCINSKI (*Member*)
Grainger Professor of Nuclear Engineering, University of Wisconsin

GERALD KULCINSKI is the Grainger Professor of Nuclear Engineering and Director of the Fusion Technology Institute at the University of Wisconsin. He received a B.S. degree in chemical engineering and M.S. and Ph.D. degrees in nuclear engineering from the University of Wisconsin. He has held research positions at Los Alamos Scientific Laboratories and Battelle Northwest Laboratory involving nuclear rockets and radiation damage reactor materials. Dr. Kulcinski has served as a consultant to various industries and on many technical advisory committees. He is a member or fellow of several scientific societies. His current research activities are focused on magnetic and inertial confinement fusion.

Professor Kulcinski is not currently performing any SDI-related research. From 1986 through October 1987 he participated in an Air Force assessment of the potential for using fusion power in space. He does energy-related research sponsored by DOE, USAF, and NASA; all of that work is related to terrestrial power. In November 1987 his department at the University of Wisconsin received a subcontract from Ebasco Corporation to engage in an SDIO study on SMES. Professor Kulcinski has not participated in that study.

JEROME P. MULLIN (*Member*)
Vice President, Research, Advanced Technology Group, Sundstrand Corporation

JEROME MULLIN is Vice President, Research, Advanced Technology Group, with the Sundstrand Corporation. He received a B.S. degree in physics from Spring Hill College, graduate training from the University of Maryland, and topical training from Princeton, Arizona State, and Catholic University. In addition, he was a 1973 Stanford Sloan Fellow. Mr. Mullin was previously with NASA as a program manager, systems program manager, and director of its Space Energy Systems Office. From 1963 to 1984 he directed NASA's space power research program. Among other honors, he received the Exceptional Service Medal of the National Aeronautics and Space Administration.

At Sundstrand, Mr. Mullin is research director for the Advanced Technology Group in areas relevant to corporate products for aircraft and spacecraft subsystems. From 1963 to 1984, he directed NASA's space power research program, which overlapped areas of research covered in this study, such as SP-100, solar systems, thermal management, and power conditioning. Mr. Mullin is not personally engaged in his group's SDIO research. Sundstrand has participated as a subcontractor in a number of SDIO studies, including one of the Space Power Architecture System (SPAS) studies, the concept and requirements definition (CARDS) study for a neutral particle beam platform, and a megawatt-burst thermal management study. Sundstrand expects to be a subcontractor on multimegawatt advanced Rankine technology, and is presently a supporting participant in a Stirling engine proposal to NASA. Sundstrand has also been active in the development of organic Rankine cycle technology. Isotope-fueled versions of this technology are under study for the Boost Surveillance and Tracking System (BSTS), and solar-powered versions have been evaluated for the Space Station. Sundstrand expects to pursue similar activities in the future. The totality of SDIO-related activities of its Advanced Technology Group make up less than one percent of its total effort.

K. LEE PEDDICORD (*Member*)
Professor of Nuclear Engineering, Texas A&M University

K. LEE PEDDICORD is a Professor of Nuclear Engineering at Texas A&M University and Assistant Director for Research at the Texas

Engineering Experiment Station. He received a B.S. degree in mechanical engineering from the University of Notre Dame and M.S. and Ph.D. degrees from the University of Illinois. Dr. Peddicord has been active in research and education in the field of nuclear engineering and nuclear fission. He has held positions at the Swiss Federal Institute for Reactor Research, Oregon State University, the EURATOM Joint Research Centre, and was formerly head of the Department of Nuclear Engineering at Texas A&M University. He is a member of several professional and educational societies, and an author or coauthor of over 100 scientific publications and reports.

Professor Peddicord is a consultant to Los Alamos National Laboratory, including serving on a materials science review committee for the SP-100 project office, and is engaged in or supervises other DOE-sponsored research, including projects relating to multimegawatt nuclear fuel behavior. He has served on the Materials Science Review Committee for the SP-100 program.

CAROLYN K. PURVIS (*Member*)
Chief, Spacecraft Environment Office, NASA Lewis Research Center

CAROLYN PURVIS is Chief of NASA Lewis Research Center's Spacecraft Environment Office, which conducts investigations to define and evaluate interactions between space systems and orbital environments. She holds a B.A. degree in physics from Cornell University, an M.S. degree in physics from the university of Washington, and a Ph.D. degree in theoretical solid-state physics from Case Western Reserve University. Dr. Purvis is a member of the American Geophysical Union and the American Physical Society, and is an Associate Fellow of the American Institute of Aeronautics and Astronautics. She is the author or coauthor of over 35 technical publications and reports.

Dr. Purvis manages an investigation to define and develop analytical tools to evaluate space environment effects on SDIO systems. This work comprises both in-house activity and the management of outside study grants and contracts. She is a member of an independent DNA review committee for the SDI Space Plasma Experiments Aboard Rockets (SPEAR) program. The Spacecraft Environment Office is also conducting some environment compatibility studies in support of the NASA/SDIO SP-100 Advanced Development Program. That activity is managed by another NASA Lewis group.

WALTER J. SARJEANT (*Member*)
Director of the Power and Power Conditioning Institute and Professor of Electrical Engineering, State University of New York at Buffalo

WALTER SARJEANT is Director of the Space Power and Power Conditioning Institute and the James Clerk Maxwell Professor of Electrical Engineering at the State University of New York at Buffalo. He received BSc., MSc., and Ph.D. degrees from the University of Western Ontario in the field of electric discharge lasers. He has been a member of or directed various research, design, and development groups within national laboratories and industry in the areas of pulse power components and impulse measurement systems. At Los Alamos National Laboratory he was responsible for advanced development of pulse power components and systems, including energy storage devices, switches, and high-power modulators. Dr. Sarjeant has led government-sponsored research projects and technical advisory committees in the field of power and power conditioning. He is a senior member of the Institute of Electrical and Electronics Engineers and is a member of the New York Academy of Sciences.

Professor Sarjeant is doing research on an SDIO contract via DNA on insulation for space power applications. He is chairman of an independent DNA review committee for the SDI SPEAR program. Through W. J. Schafer Associates, Inc., Dr. Sarjeant serves as a technical adviser to DNA on the advanced simulator program and on DNA-managed SDIO power programs. He is a consultant to the Los Alamos National Laboratory and to Sandia National Laboratories in the fields of power, lasers, and accelerators.

J. PACE VANDEVENDER (*Member*)
Director, Pulsed Power Sciences, Sandia National Laboratories

PACE VANDEVENDER is Director of Pulsed Power Sciences at Sandia National Laboratories, Albuquerque, New Mexico. He has a B.A. degree in physics from Vanderbilt University, an M.A. degree in physics from Dartmouth College, and a Ph.D. degree in physics from the Imperial College of Science & Technology, University of London, England. He has managed or supervised pulsed power or fusion research projects at Sandia since 1974. His present work covers pulsed power R&D, inertial confinement fusion, nuclear weapons effects simulation, and kinetic and directed-energy-weapon R&D for tactical applications and for the SDI. He serves on several advisory committees at Sandia National Laboratories, Los Alamos National Laboratory, and the SDIO. Dr. VanDevender has received various academic honors, and in 1984 was named one of "100 Most Promising Scientists Under 40" by *Science Digest.*

Dr. VanDevender has served on advisory panels to SDIO in the area of directed energy. Personnel in his directorate have managed projects and conducted in-house work on power conditioning for the SDIO Power Program Office. Directorate personnel also work on electron beam technology, lasers, and electromagnetic launchers for other programs within SDIO. Altogether, SDIO provides approximately 15 percent of the funding in his organization.

ENERGY ENGINEERING BOARD LIAISON

S. WILLIAM GOUSE (*Energy Engineering Board Liaison*)
Senior Vice President and General Manager, Civil Systems Division, MITRE Corporation

WILLIAM GOUSE is Senior Vice President and General Manager of the Civil Systems Division of the MITRE Corporation in McLean, Virginia. He holds B.S., M.S., and Sc.D. degrees in mechanical engineering from MIT. Dr. Gouse has held academic positions at MIT and Carnegie-Mellon University. He served as Technical Assistant for Civilian Technology in the Office of Science and Technology, Executive Office of the President; Science Advisor to the Secretary of Interior; Director of the Office of Coal Research; and Deputy Assistant Administrator for Fossil Energy in the Energy Research and Development Administration. Dr. Gouse is a member of many scientific and educational societies, has served as a consultant to industry, and has participated on various National Academy of Sciences/National Research Council study committees. He is a member of the Energy Engineering Board.

Dr. Gouse is not personally engaged in SDI-related research. His organization, the Civil Systems Division of The MITRE Corporation, is also not engaged in SDI work. The C^3 Air Force Group of The MITRE Corporation in Bedford, Massachusetts, conducts some work in support of SDI, amounting to about 5 percent of its activity.

TECHNICAL ADVISOR

Z. J. JOHN STEKLY
Vice President, Advanced Programs, Intermagnetics General Corporation

JOHN STEKLY is Vice President, Advanced Programs, of Intermagnetics General Corporation. He studied electrical and mechanical engineering at MIT, where he received B.S., M.S., and Ph.D. degrees. Dr. Stekly has held management and research positions at various corporations in the field of applied superconductivity, and in areas relating to utility, military, and medical applications. He is a member of the National Academy of Engineering, and has served on National Academy of Sciences/National Research Council committees on magnetic fusion, electrical energy, and materials for the Army.

Intermagnetics General Corporation, a small business, is a major developer and manufacturer of superconducting materials, superconducting and permanent magnets, and cryogenic systems. The company has several SDI-funded development contracts or subcontracts in conductors, permanent magnets, refrigeration, and superconductivity applications. Dr. Stekly periodically consults on these efforts.

STUDY DIRECTOR

ROBERT COHEN
Senior Program Officer, Energy Engineering Board, National Academy of Sciences

ROBERT COHEN is a Senior Program Officer at the Energy Engineering Board, National Research Council of the National Academy of Sciences. He has a B.S. degree in chemistry from Wayne State University, an M.S. degree in physics from the University of Michigan, and a Ph.D. degree in electrical engineering from Cornell University. Dr. Cohen's first career was research on remote sensing of plasma phenomena in the ionosphere, utilizing radio and radar techniques, in Colorado and South America for the National Bureau of Standards and later the National Oceanic and Atmospheric Administration. In 1973 he initiated the federal ocean energy R&D program. He is a member of the IEEE Energy Committee.

APPENDIX

C
Study Chronology
(Meetings, Briefings, and Site Visits)

MEETING, APRIL 21–22, 1987, WASHINGTON, D.C., NATIONAL ACADEMY OF SCIENCES

April 21, 1987

Overview of the Independent Evaluation Group (IEG) for the SDIO power program	R. Joseph Sovie NASA Lewis Research Center
Review of the SP-100 Ground Engineering System (GES) space reactor program	Major Joseph A. Sholtis U.S. Department of Energy
Overview of the multimegawatt space nuclear power program	Stephen J. Lanes U.S. Department of Energy

Classified presentations:

Philosophy of the SDIO space-based power program; relevance of this study	Richard Verga and Robert Wiley SDIO Power Program Office
Overview of the SDI system architecture	Edward T. Gerry W. J. Schafer Associates, Inc.
Overview of the SDI power program	Richard Verga and Robert Wiley SDIO Power Program Office

After-dinner talk:

Possible civilian space-based power applications	Raymond S. Colladay NASA Headquarters

April 22, 1987: Committee Executive Session

MEETING, JUNE 25–26, 1987, NASA LEWIS RESEARCH CENTER, CLEVELAND, OHIO

June 25, 1987

NASA space power needs and programs	R. Joseph Sovie NASA Lewis Research Center
SDI space power architecture studies	R. Joseph Sovie NASA Lewis Research Center
SDI nonnuclear baseload and multimegawatt power program	William Borger Air Force Wright Aeronautical Laboratory
SDI power conditioning and pulse power program	Phillip N. Mace W. J. Schafer Associates, Inc.
Space plasma experiments aboard rockets (SPEAR)	Herbert Cohen W. J. Schafer Associates, Inc.
USAF/SDI thermionics technology program	Elliot Kennel Air Force Wright Aeronautical Laboratory
Space power environmental effects	Carolyn K. Purvis Committee on Advanced Space Based High Power Technologies

After-dinner talk:

Powering the space station	Larry H. Gordon NASA Lewis Research Center

June 26, 1987: Executive Session

Classified presentation:

Soviet space nuclear power program	Lt. Randy Wharton USAF/Wright Patterson

MEETING, JULY 20–21, 1987, ALBUQUERQUE, NEW MEXICO

July 20, 1987: Sandia National Laboratories

Issues in nuclear fuel technology for space power under transient conditions	K. Lee Peddicord Committee on Advanced Space Based High Power Technologies
Status of multimegawatt power sources and power conditioning subsystems	Louis O. Cropp Sandia National Laboratories
NPB integrated experiment power subsystem integration	Quentin Quinn McDonnell Douglas
The DOE thermionic fuel element (TFE) verification program	Richard Dahlberg General Atomics

Classified presentations:

Power system aspects of the neutral-particle beam (NPB)	Carmelo Spirio Los Alamos National Laboratory
An Air Force perspective on space power	Capt. Arthur F. Huber AFSTC/XL (PLANS) Kirtland Air Force Base New Mexico
Soviet space-based power development	Phil Berman Defense Intelligence Agency

After-dinner talk:

High-temperature superconductors: Why all the excitement?	Frederic A. Morse Los Alamos National Laboratory

July 21, 1987 (a.m.): Sandia National Laboratories

Tour of PBFA II	Guide: T. H. Martin Sandia National Laboratories
Tour of Sandia Pulsed Reactor III	Guide: K. R. Prestwich Sandia National Laboratories

July 21, 1987 (p.m.): Executive Session at Offices of Science Applications International Corporation

MEETING, AUGUST 25–26, 1987, SEATTLE, WASHINGTON

August 25, 1987/Boeing Aerospace Company

Overview of the free-electron laser (FEL)	Don R. Shoffstall Boeing Aerospace Co.
Tour of the FEL laboratory	Guide: Don R. Shoffstall Boeing Aerospace Co.
Beaming power from earth to space with microwaves	Walter S. Gregorwich Lockheed Missiles & Space Co.
U.S. Army activities relevant to space-based power	Larry I. Amstutz Belvoir Research, Development, and Engineering Center
The SP-100 project's approach to developing a long-lived SP-100 power system	Jack F. Mondt Jet Propulsion Laboratory

August 26, 1987: Executive session at Battelle Seattle Conference Center

MEETING, OCTOBER 19–20, 1987, NATIONAL ACADEMY OF SCIENCES, WASHINGTON, D.C.

October 19, 1987

Update on SDIO power program	Richard Verga and David Buden SDIO Power Program Office
Presentation on beaming of power to space platforms using microwaves	William C. Brown Microwave Power Transmission Systems and Martin I. Hoffert New York University

October 20, 1987

Status of Congressional activities relevant to space power	John V. Dugan and Nelson L. Milder Subcommittee on Energy R&D House Committee on Science, Space, and Technology

MEETING, NOVEMBER 17–18, 1987, NATIONAL ACADEMY OF SCIENCES WASHINGTON, D.C.

November 17, 1987

Some innovative concepts for space power generation and storage	Osman K. Mawardi Case Western Reserve University
Study panel on small nuclear space reactor power systems	Earl J. Wahlquist U.S. Department of Energy

November 18, 1987: Executive Session

MEETING, JANUARY 21–22, 1988, NATIONAL ACADEMY OF SCIENCES, WASHINGTON, D.C.

January 21, 1988

Report on Albuquerque meeting on space nuclear power	Gerald L. Kulcinski and K. Lee Peddicord Committee on Advanced Space Based High Power Technologies

January 22, 1988

Update on the SDIO power program	Richard Verga SDIO Power Program Office

APPENDIX

D Possible Impacts of Effluents from SDI Systems

SPACE SHUTTLE EXPERIENCE RELEVANT TO POSSIBLE IMPACTS OF EFFLUENTS PROJECTED FOR SDI SYSTEMS

Space Shuttle experience is relevant to projecting likely effects of effluent dump rates for Strategic Defense Initiative (SDI) systems. A good source of data on effluent rates for the Space Shuttle was provided by Pickett et al. (1985). Sources of effluent on the Shuttle include outgasing (estimated at about 3×10^{-4} kg/s at mid-mission), Flash Evaporator System (FES) operations, water dumps, and the primary and vernier Reaction Control System (RCS) engines. Data found for the STS-3 mission are presumed here to be typical of those for other Space Shuttle missions.

The FES system dumps water vapor in pulses of duration 200 $\pm$ 30 ms at a maximum pulse rate of 4 Hz, yielding a release rate of 22.7 kg/h (= 6.3×10^{-3} kg/s). The average FES release rate is 2.3 kg/h (= 6.3×10^{-4} kg/s) at 0.4 Hz. There were 20 FES releases during the STS-3 mission, with durations ranging from 1 min to about 2.5 h.

The Space Shuttle also dumps liquid water at an average rate of 64 kg/h (= 1.8×10^{-2} kg/s). During the STS-3 mission, there were nine water dumps, each of duration 45–60 min, which released a total of 41–93 kg of water. If the 93 kg is assumed to have been dumped in 60 min, this gives a maximum rate of 2.6×10^{-2} kg/s.

The RCS consists of 38 primary thrusters (395 kg of thrust each) and 6 vernier thrusters (11 kg of thrust each). These thrusters use monomethyl hydrazine (MMH) fuel with an NO_3 oxidizer, and have (calculated) effluent as shown in Table D-1 (Pickett et al., 1985, Table 2). MMH is $N_2H_3CH_3$. Minimum pulse duration is 80 ms. Longest pulse on STS-3 was about 30 s. Mass efflux rate for the vernier thrusters is 4×10^{-2} kg/s per engine, and 1.4 kg/s per engine for the primary thrusters. The velocity of the released gases is estimated as 3.5 km/s.

Finding estimates for SDI system effluents proved somewhat more difficult. Review of the Space Power Architecture System (SPAS) summary reports (1988) yielded a quotation from TRW, which performed one of the three SPAS studies, of 7.5 kg/s of H_2 for cooling a 180-MW free-electron laser (FEL), and 13.5 kg/s of H_2 for cooling a 400-MW neutral-particle beam (NPB). Martin-Marietta gives estimates in the 40–100 kg/s range for various NPB systems (Tables D-2 and D-3), and General Electric (GE) quotes efflux rates in the 20 to 40 kg/s range, though these are associated with power systems, so this may amount to comparing apples with oranges (Table D-4). The wide variation in these numbers is due to different assumptions about power levels and system designs among the contractors.

It appears that, with the exception of the primary RCS thrusters, none of the efflux rates associated with the Shuttle approach those estimated for SDI systems. Neither is there much H_2 efflux from the Shuttle. The L2U burn (described in the following section), during which the three PRCS engines were fired over a period of 1.5 min, was the longest PRCS operation identified by the committee. Data from this burn have been used in the following section to estimate the usefulness of the spherical assumption employed for efflux expansion.

ESTIMATION OF THE IMPACT OF EFFLUENT ON PROPAGATION OF A NEUTRAL-PARTICLE BEAM

Following is a calculation—using a spherical approximation—to estimate the impact of effluent on propagation of a neutral-particle beam. Assume that the effluent originates at the origin and expands radially with velocity $u\hat{r}$. The resulting mass density ρ at distance R is given by

TABLE D-1 Thruster Plume Characteristics for Primary (PRCS) and Vernier (VRCS) Thrusters

Effluent Species	Molecular Weight	Mole Fraction
	Composition, Neutrals	
H_2O	18	0.328
N_2	28	0.306
CO_2	44	0.036
O_2	32	0.0004
CO	28	0.134
H_2	2	0.17
H	1	0.015
$MMH\text{-}NO_3$	108	0.002
Total		0.9914
	Composition, Dominant Ions	
NO^-	30	1.7×10^{-8}
CO_2^-	44	2.7×10^{-10}
OH^-	17	4.3×10^{-10}
Electrons	--	24×10^{-9}

Thruster Firing	Number of Neutrals Ejected	Number of Ions (electrons) Ejected
VRCS		
Typical[a]	1.3×10^{25}	3.1×10^{17}
Longest[b]	1.7×10^{26}	3.8×10^{18}
PRCS		
Typical[c]	9.2×10^{24}	2.1×10^{17}
Longest[d]	5.5×10^{25}	1.2×10^{18}

NOTE: For the primary thruster, $\dot{m}$ = 1,419.8 g/s/engine, where $\dot{m}$ is the mass efflux rate; for the vernier thruster, m = 40.8 g/s/engine. PRCS = primary reaction control system; VRCS = vernier reaction control system. MMH = monomethylhydrazine.

[a] Based on 2 firings ejecting 163 g over 2 s.
[b] Based on 14 firings ejecting 2,100 g over 30 s.
[c] Based on 1 firing ejecting 114 g over 80 ms.
[d] Based on 5 firings ejecting 682 g over 720 ms.

SOURCE: Pickett et al. (1985): Table 2.

TABLE D-2 Assumed Effluent Compositions

Power System	Effluent Compositions in Weight Percent
Nuclear Brayton turboalternator	100% H
Nuclear Rankine turboalternator	No effluent
Liquid metal reactor	No effluent
Combustion Brayton turboalternator	FEL: 52.8% H_2, 47.2% H_2O EML/NPB: 58.6% H_2, 41.4% H_2O
Combustion Brayton turboalternator (water collected)	100% H_2
Combustion-driven MHD	FEL: 20.2% H_2, 66.6% H_2O, 4.2% CsOH[a] EML/NPB: 36.1% H_2, 60.1% H_2O, 3.8% CsOH[a]
Fuel cell	FEL: 59.1% H_2, 40.9% H_2O EML/NPB: 67.6% H_2, 32.4% H_2O
Fuel cell, radiator cooled	No effluent
Fuel cell, water collected	100% H_2

NOTE: EML = electromagnetic launcher; FEL = free-electron laser; MHD = magnetohydrodynamics; NPB = neutral-particle beam.

[a]Actual cesium-containing species may differ from this.

SOURCE: Martin-Marietta Space Power Architecture System (1988) report, Task 3: Table III-2.1.

$$\rho(R) = \frac{\dot{m}}{4\pi R^2 u}, \tag{1}$$

where $\dot{m}$ is the mass flow rate. In propagating from R_o to a target at R_f, the neutral-particle beam traverses a mass per square centimeter given by

$$< \rho R >= \int_{R_o}^{R_f} \rho dR = \frac{\dot{m}}{4\pi u}\left(\frac{1}{R_o} - \frac{1}{R_f}\right),$$

which, for $t \gg R_o/u$ and $R_f \gg R_o$, becomes

$$< \rho R >= \frac{\dot{m}}{4\pi u R_o}. \tag{2}$$

For hydrogen exhaust at 1000°K, $u \approx 4.5 \times 10^3$ m/s. The cross section for 100 MeV H_o on H_o is 1.23×10^{-19} cm^2 in most recent work (Johnstone, 1988). The corresponding integrated mass density of effluent required to ionize 50 percent of the neutral-particle beam is 8.9×10^{-6} g/cm^2. Approximately 7 percent of the beam is stripped

TABLE D-3 Assumed Effluent Initial Temperatures (°K), Initial Pressures (atm), and Flow Rates (kg/s)

		FEL		NPB		EML	
	Pressure	Temp.	Flow	Temp.	Flow	Temp.	Flow
Nuclear Brayton turboalternator	1.0	586	52	416	44	855	44
Nuclear Rankine turboalternator	--	--	--	--	--	--	--
Liquid metal reactor	--	--	--	--	--	--	--
Combustion Brayton	1.0	570	89	407	70	818	70
Combustion Brayton, water collected	1.0	639	48	447	41	996	41
Combustion-driven MHD	1.5	3000	178	3000	122	3000	122
Fuel cell	1.0	450	81.7	450	61.8	450	61.8
Fuel cell, radiator cooled	--	--	--	--	--	--	--
Fuel cell, water collected	1.0	450	48.3	450	41.8	450	41.8

NOTE: EML = electromagnetic launcher; FEL = free-electron laser; MHD = magnetohydrodynamics; NPB = neutral-particle beam.

SOURCE: Martin-Marietta Space Power Architecture System (1988) report, Task 3: Table III-2.2.

at 1×10^{-6} g/cm². For purposes of the present calculation, the integrated mass density for which beam stripping becomes a concern is taken to be

$$< \rho R >= 10^{-6} g/cm^2 = 10^{-5} kg/m^2.$$

Ensuring undisturbed beam propagation requires

$$\dot{m} \ll 4\pi u R_o < \rho R >= 0.56 R_o kg/s,$$

or, for $R_o = 10$ m, $\dot{m} \ll 5.6$ kg/s. As noted above, estimates of coolant mass efflux rates for neutral-particle beams used in the Space Power Architecture System (SPAS) studies were in the range of about 10 to 100 kg/s, somewhat greater than the 5.6 kg/s estimated above. Thus, if the spherical approximation is at all reasonable, neutral-particle beam stripping should be a serious concern.

A zero-order check on the spherical approximation can be made based on Space Shuttle data from the STS-3 flight. During this flight, an engine test of the Shuttle's Primary Reaction Control System (PRCS) was conducted. Observations of pressure in the bay

TABLE D-4 Exhaust Properties

Power System	Mass Flow (kg/MWe·s)	Total Energy (MWe)	Total Mass Flow (kg/s)	Species	Mole Fraction	Mass Fraction	Total Number Density (#/cm^2·s)	Total Flux Density (#/cm^2·s)
TM-1	0.20	200	39.6	H_2	0.916	0.549	2.4×10^{17}	8.5×10^{22}
				H_2O	0.084	0.450	2.2×10^{16}	7.7×10^{21}
TM-4	0.09	400	37.7	H_2	1.0	1.0	2.8×10^{17}	1.2×10^{23}
MHD	0.12	200	24.3	H_2	0.995	0.904	6.6×10^{16}	4.5×10^{22}
				Cs	0.002	0.094	1.3×10^{14}	8.9×10^{19}
				H	0.003	0.001	1.9×10^{14}	1.3×10^{20}

NOTE: MHD = magnetohydrodyamics; "#" = number of molecular particles.

SOURCE: General Electric Space Power Architecture System (1988) report, Task 4: Table 3.

were made by the neutral pressure gauge on the Plasma Diagnostics Package (PDP) during this engine test, known as the L2U burn. This burn lasted for about 1.5 min, and involved pulsed firing of the L2U and R1U thrusters and continuous firing of the F2U thruster.

Pressure at the PDP's location was about 3×10^{-4} torr (compared to a background pressure of 10^{-7} torr) during the L2U burn (Murphy et al., 1983; Pickett et al., 1985; Shawhan and Murphy, 1983). The mass flow rate of each of the PRCS engines is $\dot{m} = 1.42$ kg/s (Murphy et al., 1983, Pickett et al., 1985; Table D-1), which is high enough that the exhaust gas is collisional when it exits, as expected for coolant efflux associated with SDI systems. The L2U and R2U thrusters are located to the port and starboard, respectively, of the Shuttle's vertical stabilizer on the orbital maneuvering system (OMS) pods and fire upward, while the F2U engine is on the upper surface of the Shuttle's nose and also fires upward (Murphy et al., 1983). During the L2U burn, the engines were fired to obtain net zero torque on the Shuttle, with R1U cancelling the roll induced by L2U and F2U cancelling pitch from both rear thrusters. The PDP is located on-axis, somewhat aft of the center of the payload bay, at about the level of the longerons.

Estimates of the distances from the PDP to the three engines are as follows:

- F2U: 17 m forward, on-axis, 1.5 m down;
- R1U: 12 m aft, 2.7 m to starboard, 1.5 m up; and
- L2U: 12 m aft, 2.7 m to port, 1.5 m up.

These estimates yield distances of $R_F \approx 17$ m to F2U and $R_R \approx 13$ m to R1U and L2U. The mass density at R_o, the location of the pressure gauge, is computed using equation (1) as the sum of contributions from the three engines. The result for mass density is:

$$\rho(R_o) = \frac{\dot{m}}{4\pi u}\left(\frac{2}{R_R^2} + \frac{1}{R_F^2}\right). \tag{3}$$

To relate this result to pressure, recall that

$$p = \frac{1}{3}\rho \bar{v}^2, \tag{4}$$

where $\bar{v} = \sqrt{2kT/m}$ and T is the temperature of the gas at the point in question. An upper limit on the pressure can be obtained by assuming that $\bar{v} = u$; that is, there is no cooling of the effluent. Making this assumption, and using $u = 3.5$ km/s (Pickett et al., 1985) yields, for the Shuttle case above,

$$p_{max}(R_o) \approx 2.1 \text{ nt/m}^2 \approx 1.6 \times 10^{-2} \text{torr}.$$

A reasonable lower bound on the expected pressure may be obtained by assuming that the effluent thermalizes; that is, $T \approx 200°K–300°K$ at the pressure gauge. The velocity $u = 3.5$ km/s corresponds to an average kinetic temperature of 22000°K for the PRCS exhaust (Murphy et al., 1983; Pickett et al., 1985). Using this correspondence, and taking $T = 250°K$ for the thermalized exhaust, yields

$$p_{thermal} \approx 1.8 \times 10^{-4} \text{torr}.$$

The measured pressure for the L2U burn was 3×10^{-4} torr, which is very close to the pressure computed above for the thermalized case. This suggests that the spherical approximation is reasonable, which in turn suggests that neutral particle beam stripping by effluent is an issue that must be addressed.

The committee considers that a flight experiment to characterize the behavior of the H_2 effluent would be appropriate. Altitude is an important consideration here. At Shuttle altitudes, orbital vehicles move at about 7.5 km/s, and the residual atmosphere is dense enough to cause rapid dissipation of effluent clouds. At higher altitudes, the residual atmosphere becomes very tenuous, so the effluent will remain near the vehicle longer. To the extent that the orbits of interest are in the 1,000+ km range, useful early data might be obtained from a sounding rocket experiment, because sounding rocket velocities relative to the background residual atmosphere are on the same order as the random thermal velocity of the background atoms (about 1 km/s). Even though the residual atmospheric densities are high at typical sounding rocket altitudes of a few hundred kilometers, the low relative velocity of the rocket should make the dissipation of rocket-generated effluent comparable to that expected of vehicles orbiting at higher altitude (greater than 1,000 km).

APPENDIX

E
Compilation of Study Findings, Conclusions, and Recommendations

Following is a compilation of the study findings, conclusions, and recommendations.

FINDINGS

The committee arrived at the following findings:

Finding 1 (Chapter 2): Of the three significantly different SDI modes of operation (housekeeping, alert, and burst mode), requirements for the alert mode are inadequately defined, yet they appear to be a major design-determinant. For that mode, the unprecedented high power levels, durations, and unusual time-profiles–as well as the associated voltages and currents–which are envisioned will usually make extrapolation from previous experience quite risky and unreliable. A possible exception is in the area of turbine technology, where an adequate range of power levels has been validated for terrestrial applications, although not for flight conditions. Proposed space power systems will need to be space qualified for long-term unattended use.

Finding 2 (Chapter 4): The space power subsystems required to power each SDI spacecraft are a significant part of a larger, complex system into which they must be integrated, hence for obtaining a valid analysis they cannot be treated completely separately. (Se Conclusion 2 and Recommendation 1.)

Finding 3 (Chapter 4): Existing space power architecture sys tem studies do not adequately address questions of survivabilit reliability, maintainability, and operational readiness—that is, avai ability on very short notice.

Finding 4 (Chapter 4): Existing SDI space power architectur system studies do not provide an adequate basis for evaluating o comparing cost or cost-effectiveness among the space power system examined.

Finding 5 (Chapter 2): Among the power systems that ar candidates for SDI applications, the least massive, autonomous self contained space power systems currently being considered entail tol erance of substantial amounts of effluent during system operation The feasibility of satisfactorily operating spacecraft sensors, weapons and power systems in the presence of effluent is still unresolved.

Finding 6 (Chapter 3): Beaming power upward from earth b microwaves or lasers (see Recommendation 6) has not been exten sively explored as a power or propulsion option.

Finding 7 (Chapter 6): The present overall rate of progres in improving the capability of space power-conversion and power conditioning components appears inadequate to meet SDI schedule or NASA needs beyond the Space Station.

Finding 8 (Chapter 3): The time needed for the developmen and demonstration of a U.S. space nuclear reactor power systen currently exceeds the time required to plan and deploy a missio dependent upon that power source.

CONCLUSIONS

The committee reached the following conclusions:

Conclusion 1 (Chapter 2): Multimegawatt space power sources (at levels of tens to hundreds of megawatts and beyond) will be a necessity if the SDI program is to deploy electrically energize weapons systems for ballistic missile defense.

Conclusion 2 (Chapter 4): Gross estimated masses of SDI space ›ower systems analyzed in existing studies appear unacceptably large o operate major space-based weapons. At these projected masses, he feasibility of space power systems needed for high-power SDI con-epts appears impractical from both cost and launch considerations. \venues available to reduce power system costs and launch weights nclude (a) to substantially reduce SDI power requirements; (b) to ignificantly advance space power technology.

Conclusion 3 (Chapter 3): The amount of effluent tolerable is a :ritical discriminator in the ultimate selection of an SDI space power ;ystem. Pending resolution of effluent tolerability, open-cycle power ;ystems appear to be the most mass-effective solution to burst-mode ›lectrical power needs in the multimegawatt regime. If an open-cycle ;ystem cannot be developed, or if its interactions with the spacecraft, weapons, and sensors prove unacceptable, the entire SDI concept will be severely penalized from the standpoints of cost and launch weight (absent one of the avenues stated in Conclusion 2, Chapter 4).

Conclusion 4 (Chapter 2): The rate of rise ("ramp-rate") from zero to full burst-mode power level appears to be a critical requirement. It is not apparent to the committee what relationships exist among elapsed time for power build-up and system complexity, mass, cost, and reliability.

Conclusion 5 (Chapter 5): Major advances in materials, components, and power system technology will be determining factors in making SDI space power systems viable. Achieving such advances will require skills, time, money, and significant technological innovation. The development of adequate power supplies may well pace the entire SDI program.

Conclusion 6 (Chapter 6): Refocusing SDIO resources toward near-term demonstrations could delay development of advanced power technology, and thereby seriously jeopardize meeting long-term space power program objectives.

Conclusion 7 (Chapter 3): A space nuclear reactor power system, once available, could serve a number of applications—for example, in NASA and military missions requiring up to 100 kWe of power or more—in addition to SDI.

Conclusion 8 (Chapter 2): Survivability and vulnerability concerns for SDI space power systems have not yet been adequately addressed in presently available studies relevant to SDI space power needs.

RECOMMENDATIONS

The committee arrived at the following recommendations:

Recommendation 1 (Chapter 4): **Using the latest available information, an in-depth full-vehicle-system preliminary design study–for two substantially different candidate power systems for a common weapon platform–should be performed now, in order to reveal secondary or tertiary requirements and limitations in the technology base which are not readily apparent in the current space power architecture system studies.** Care should be exercised in establishing viable technical assumptions and performance requirements, including survivability, maintainability, availability, ramp-rate, voltage, current, torque, effluents, and so on. This study should carefully define the available technologies, their deficiencies, and high-leverage areas where investment will produce significant improvement. The requirement for both alert-mode and burst-mode power and energy must be better defined. Such an in-depth system study will improve the basis for power system selection, and could also be helpful in refining mission requirements.

Recommendation 2 (Chapter 3): **To remove a major obstacle to achieving SDI burst-mode objectives, estimate as soon as practicable the tolerable on-orbit concentrations of effluents.** These estimates should be based—to the maximum extent possible—on the results of space experiments, and should take into account impacts of effluents on high-voltage insulation, space-platform sensors and weapons, the orbital environment, and power generation and distribution.

Recommendation 3: Rearrange space power R&D priorities as follows:

a. **(Chapter 3) Give early, careful consideration to the regulatory, safety, and National Environmental Policy Act requirements for space nuclear power systems from manufacture through launch, orbital service, safe orbit requirements, and disposition.**

b. **(Chapter 6) The SP-100 nuclear power system is applicable both to SDI requirements and to other civil and military space missions. Therefore SP-100 development should be completed, following critical reviews of SP-100 performance goals, design, and design margins.**
c. **(Chapter 6) SDI burst-mode requirements exceed by one or more orders of magnitude the maximum power output of the SP-100. Therefore both the nuclear and nonnuclear SDI multimegawatt programs should be pursued.** Hardware development should be coordinated with the results of implementing Recommendation 5.

Recommendation 4 (Chapter 6): Consider deploying the SP-100 or a chemical power system on an unmanned orbital platform at an early date. Such an orbital "wall socket" could power a number of scientific and engineering experiments. It would concurrently provide experience relevant to practical operation of a space power system similar to systems that might be required by the SDI alert and burst modes.

Recommendation 5 (Chapter 5): Make additional and effective investments now in technology and demonstrations leading to advanced components, including but not limited to:

- **thermal management, including radiators;**
- **materials—structural, thermal, environmental, and superconducting;**
- **electrical generation, conditioning, switching, transmission, and storage; and**
- **long-term cryostorage of H_2 and O_2.**

Advances in these areas will reduce power system mass and environmental impacts, improve system reliability, and, in the long term, reduce life-cycle power system cost.

Recommendation 6 (Chapter 3): Review again the potential for ground-based power generation (or energy storage) with subsequent electromagnetic transmission to orbit.

Recommendation 7 (Chapter 2): After adequate evaluation of potential threats, further analyze the subject of vulnerability and survivability, mainly at the overall system level. Data resulting from implementing Recommendation 1 would be appropriate for this analysis. Pending such analysis, candidate power systems should be screened for their potential to satisfy interim SDIO survivability requirements, reserving judgment as to when or whether those requirements should constrain technology development. Convey the screening results to the advocates of those candidate power systems, to stimulate their finding ways to enhance survivability as they develop the technology.

Recommendation 8 (Chapter 6): To further U.S. capabilities and progress in civil as well as military applications of power technology, both on the ground and in space, and to maintain a rate of progress in advanced technologies adequate to satisfy national needs for space power, plan and implement a focused federal program to develop the requisite space power technologies and systems. This program–based on a multiyear federal commitment–should be at least as large as the present combined NASA, DOD (including SDIO), and DOE space power programs, independent of the extent to which SDI itself is funded.

Index

D

E